I0615775

Star Death

Star Death

LEO EMMANUEL LOCHARD

RESOURCE *Publications* · Eugene, Oregon

STAR DEATH

Copyright © 2010 Leo Emmanuel Lochard. All rights reserved.
Except for brief quotations in critical publications or reviews, no part
of this book may be reproduced in any manner without prior written
permission from the publisher. Write: Permissions, Wipf and Stock
Publishers, 199 W. 8th Ave., Suite 3, Eugene, OR 97401.

Resource Publications
An Imprint of Wipf and Stock Publishers
199 W. 8th Ave., Suite 3
Eugene, OR 97401
www.wipfandstock.com

ISBN 13: 978-1-60899-252-2

Manufactured in the U.S.A.

This book is dedicated to my father and my mother who by the grace of God taught us the real and true meaning of uplifting love, edifying sacrifice and faithful devotion to family.

"But the day of the Lord will come like a thief, and then the heavens will pass away with a loud noise, and the elements will be dissolved with fire, and the earth and the works that are upon it will be burned up."
—2 Peter 3:10

INTERSTELLAR MEDIUM

OUR SOLAR PLANETARY SYSTEMS DYNAMICS (SPRING)

LIFE-PLANET EARTH
23° TILT

LIFE Planet EARTH 23° TILT

PERIHELION

WINTER

SUN

AUTUMN (FALL)

COLD VOID VACCUM SPACE

APHELION

SUMMER

1

"LOOK AT A LIGHT BULB! What do you usually see? Light comes out of it. That's wonderful. We all love light especially when we need to see. But what would you say about the light bulb itself? It's good? Think for a moment! What is it made of? There is a metal filament that glows inside the glass bulb, but in a vacuum. It's glowing because it received an electrical charge, and as it glows, it releases energy in the form of photons or light rays that allow us to see; it also releases a certain amount of infrared or heat. Do you also know that the bulb is 'dying?' Say again? Yes, the bulb is in 'a state of death'—so to speak; we know, it's not alive and cannot die. But in a manner of speaking, the bulb is useful to us as it is itself 'dying.'"

"In other words, the metallic filament inside it that's glowing in a vacuum is burning out its atomic elements—shedding its mass, so to speak, in order that we might see. Specific amounts of filament mass are converted to 'excited energy' in order that the bulb might radiate light and a small amount of heat. The light radiated does not return to the bulb as 'electromagnetic mass.' It's done its job—which is, to shine. It is in an 'excited state' as the bulb glows, rather than 'congealed state' whereby it could conserve all its mass-energy. The filament returns to a 'congealed state' when we turn its light off. But a light bulb was created for shining—and that, it must do. And it does only that—for all other purposes, it is however useless. You cannot touch it or hold it lest you burn your skin, let alone eat it for dinner. That's a light bulb—it shines its light and we see, as long as the light switch is turned 'on.'"

"But light bulbs do 'burn out' or 'fade away' for one reason or another, such as in a 'power surge' that overloads the filament's capacity to glow within its specifically allowable wattage; or during a 'short-circuit,' for example, as when a positive wire meets a negative wire. And sometimes, the bulb has an increased intensity in glowing and then burns out as it 'slowly fades away.' And you take a look inside it to see where the filament 'broke off.' And if you shake it, you can sometimes hear the broken piece inside the bulb as it slams against the sides of the bulb."

"But that's not the only way light bulbs can 'go out.' Sometimes it is due to 'wear-and- tear' or 'glowing time,' or "lighting duration,' because a light bulb has a certain 'life span' or 'luminosity period' during which it can glow, expressed in 'lumen units,' 'wattage per hour' or 'life hours'—in short, the bulb's 'life's span' is expressed in how many hours of service it can provide. If the case is 'bulb-death' due to 'wear-and-tear,' the bulb just 'stops shining.' The filament may break off or it might not, depending on the circumstances that led to the faulty step causing the malfunction. There comes a point where the metal's own atoms and molecules might lose their very capacity for electro-static bonding for energy processing, flow, circulation, and release. The electrical Input-Process-Output mechanism then just stops working, analogous to when a piece of metal gets rusty whereby molecular conduction of electricity has ceased."

This conversation or soliloquy on the electro-mechanics of the light bulb took place between a father and a son, at about 5 pm on a Saturday afternoon—John Trinklung and his son Marc. John is 42 years old; Marc is 17 years old.

"Dad!" said Marc.

"Yes, Marc," replied John.

"The sun is like the biggest light bulb, isn't it?" Marc asked.

"For sure, son!" enjoined John. "You might say that, yes, in a way, yes. What made you say that?"

"O, I don't know, dad! Just thinking, I guess! They both glow and shine in different ways, so to speak!"

"Yes, that's true. Well, call Mom and Jeanie, son. It's time for supper!"

*   *   *

2

FOR THE PAST DECADE, on and off, people watching television news broadcasts and reading print magazines and newspapers have noticed certain patterns in "terra-events" or "geo-happenings" that concern the "state of the Earth"—in its ecology, bio-sphere, hydrosphere, weather systems, magnetic field, plate tectonics, etc . . . When these things touched conveniences or utilities, or supporting technologies that made human life more easily livable, then a more focused treatment was given to specific processes, because of the cause-and-effect relationships, from which, affected parties might be able to obtain satisfactory relief, in knowledge and application, equity and redress.

Since last year, the news coverage, at least for a while, focused upon farmers's complaints regarding diminished crop yields, the causes of which having not been conclusively amenable to weather activity only. Consequently, there were many areas of "agri-business,"—the term used to denote all financial and commercial activities in agriculture—that received attention in the press because of their operational importance in the "crop-raising" business. Farmers directed their complaints against fertilizer companies for degrading the quality of their products and for altering their formulae due to profit motive reasons that, they believed, were inimical to crop yields that favored revenue enrichment of farming operations. Pesticide manufacturers also received their share of farmer excoriation due to the noticeable return of certain pests, like the corn bore, which farmers had thought were definitely under control, but which proved not to be.

Corn, soybean and wheat required respective planting season treatments that were more or less standardized within the farming community as seed development took precedence over improvements in cultivation methods that were contrary to those that had stood the test of time.

That was September 2009. Since the problem with decreasing crop yields had affected the last two planting seasons, all eyes and ears were focused on the present planting season for research purposes, in order to attempt to determine the ultimate cause of these crucial changes in the agricultural industry. Feeding people was serious business. And it appeared that it was a world-wide trend with countries like Brazil, China, Russia, France and Italy, for example, reporting variations akin to gradual degradation in crop yields, as sustained in the United States of America.

Analytical and statistical evaluations from previous years have concluded that it was an event likely to continue, and that only accurate reporting procedures, research, data collection, evaluation and analysis protocols could determine a course of action to deal with the consequences, as cause-and-effect relationships had to be identified for appropriate measures and corrective actions. Soil fertility and crop yields depended upon many factors both natural and human and, since these data were routinely collected for research, development, and purposes of methods improvement, tentative results were to be provided to concerned parties before the new planting season would begin next year.

In many areas of the globe, occurrences called "dry lightning" have been taking place more often than usually expected, which was the direct cause of many forest fires, barring arson or other sources of humanly-caused disasters.

Cloud formations have been such that no rain has fallen and hence contributing to the dryness of the forest floors that caught on fire at the slightest hint of ignition. Some electricity generating power plants have reported periodic interruptions of current within their respective grids of transmission with causes not traceable to connected plant operations or to linkable environmental events. There were also transformer blowouts in certain neighborhoods, the causes of which have not been identified for accountability purposes and which required special handling due to their initial classification as "due to unknown causes."

In fact, the expression "due to unknown causes" has been gaining currency regarding a lot of problems, tragedies, and events on the Earth where human error has been eliminated as a primary or direct causal agent. No pattern of effects has emerged from preliminary investigations that would allow formation of a hypothesis to address all these developments. They appeared to have localized causes but which however have not resulted in definitively proven linkages or connections with specific individuals, groups or institutions. Every nation was taking appropriate steps in local governance, without fundamental alteration of their government systems while simultaneously engaging international organizations that focused on worldwide events for which there could be a global cause. Security was a first priority, but not at the expense of lawful, constructive liberty, especially in the presence of identifiable threats, or detectable dangers, the sources of which were still being investigated.

The incidence of skin cancers in electronic media reporting has increased in a manner that conveyed to the audience that it was widespread in industrialized nations and becoming more common in developing countries where manufacturing processes were formerly, more naturally linked to the environment. But with new economic activity based on enterprises whose business overarched the span of chemical, metallurgical and petroleum industries, smokestack emissions joined with automobile exhausts to engineer urban atmospheres analogous to those of more technologically advanced nations. However, the increases in cancer rates were occurring even in the sparsely populated countryside where the base of "making a living" was still agrarian in nature.

In highly populated mountainous areas of the globe, hospitals were reporting an unexpected increase in cardio-pulmonary disorders, including upper respiratory infections coupled with hardness of breathing or diminished ability to breathe, as if the air itself was becoming "rarefied" or losing Oxygen density. Patients reported making a great muscular effort in trying to breathe, and "taking in" so it seemed, a diminished amount of air that was incommensurate with their effort to breathe. Muscle effort was too much disproportionate to breathing ability—which caused a great influx of worrisome people into surrounding hospitals seeking alleviation of their symptoms.

In some regions of the Earth, unknown for their mild temperatures, like the deserts, a decrease in average daily temperature has been noticed, observable with an ominous absence of rain activity. In fact, from weather system logs, even many desert areas reported milder or less hot temperatures that were outside of seasonal recordings, during the summer, for example; while at the same time finding the emergence of a trend in the data, of hotter temperatures in the winter months that were uncharacteristic of weather patterns in these regions for decades past.

One farmer in Kansas reported, "the strangest thing happened on my farm"—there was a certain deformity in corn grain formations he had never seen before, an occurrence witnessed in whole rows of plant corn, while the rest of the crop stood unaffected by that husk-to-kernel deformation.

In Utah, because it had not rained for so long, the salt-content registered in Salt Lake reached a reading exceeding 50 percent whereas it normally kept steady at 12 to 13 percent salt composition.

In Texas, a rancher reported to have seen "furless mice" being fed by other mice, whose constitution was so fragile as to preclude field hunting with their peers. Another rancher swore to have seen a mature "featherless bird" that was so weak in strength as to have been a tasty meal for a coyote on its hunting foray. What was going on?—"Furless mice"? "Feather-less birds"?

In Florida, the problem was with "sink holes," the causes of which were also unknown. But then, something else was occurring, trees were sinking too, with their roots and all, with springs of water seeping or gushing out of the holes, flooding streets and unearthing sidewalks. Of course, that peninsula had always received its undeserved share of hurricanes and cyclones each year while other areas had to face tornadoes and other types of storm systems for which they also sought relief. These "forces of nature" appeared to operate all at once or not at all, as if waiting in line to strike in spurts to then remain silent for long periods of time, hence, affording researchers very few indices for putting a hypothesis together.

John Trinklung thought there had to be some kind of connection to link all these "strange" and apparently "disconnected" things happening in the past few years, and continuing into the present. He has been a farmer for nine years now, with a good knowledge of agriculture and combustion mechanics, keeping track of weather conditions, storm systems, and market news, which often required unusual alertness for systems analysis, and attention to detail in remembering seasonal associations for pattern evaluation. He has never seen "such a pattern" of events before, not at the national level, not at the global level, nor at the local level in his own neighborhood.

*   *   *

3

I T WAS 2:00 PM ON A MONDAY AFTERNOON. Suddenly the whole world "went dark"—literally, meaning, no sun, none at all. It was as if night had suddenly come, bringing with it a "pitch dark" condition all over the Earth.

A semi-truck standing at a red light advancing its bumper ever so slowly towards the tracks—just a "driving habit" of the truck driver who could no longer estimate the distance of the approaching train,—the "blasting horns" of which the truck driver could hear so well, but being not enough to prevent the accident. The inevitable occurred;—it was too late for the semi-truck's headlights to help; the train conductor could not even see the controls in order to apply the brakes. The surprise of the sun "going dark"! Too little time—things happened too fast, much too fast! They collided!

The truck "jack-knifed," the trailer separating from the cab, as the driver was ejected from the truck cabin to "find himself," landing on a flower bed's bushy shrubs nearby; miraculously he was still conscious, with only a broken leg.

The train's engine compartment at the front had somehow managed to remain on the tracks due to the angle of impact. The conductor, pushed back a few feet away from the controls and shaken by the shock, succeeded in "feeling his way" to the throttle and the brakes in order to slow the train down to a complete stop. Broken glass had scratched his arms and face as he was slightly bleeding, but not profusely so as to cause hemorrhage. Thankfully, it was a cargo train with not too many passengers on board; therefore, damage to the train was far more preferable than the loss of human lives.

Many people thought that every second ticking without sunlight was like an eternity, wondering how long the sun will remain absent from the skies.

Thirty seconds later—every thing returned to "normal." So it seemed. The sun "re-appeared" again "in the sky." It began "shining again."

But there was something very wrong with the way in which the Earth reacted—in the quickest micro-second following the sun's return, the Earth experienced the biggest "quake" or "jolt" as if it had temporarily stopped and tried to begin rotating again, or as if it was attempting to balance itself in order to regain its regular angular momentum as it was newly accelerated by solar restoration—"jerk-stop-and-go," so to speak.

An ambulance was called by the property owner on whose lawn the truck driver had landed. Land phones were functioning; cell phones could not function properly. There were discontinuities in reception and transmission of the signals, as if they were being blocked or scattered by dense electro-magnetic fields.

Some people appeared to be involved in some kind of a "dance" as they moved to and fro, seeking things to hang on to, trying to achieve equilibrium in their walk so as not to fall.

In the Amazon forests, animals that were foraging on trees fell off due to inability to move by sight or to respond to the compulsion to engage in nocturnal activities; others showed "confusion" as if being out of place for a return to nocturnal behaviors with which they were attempting to supplant diurnal activities.

In other cases, bats, perhaps perceiving that it was "night," began to fly out of their caves to then rush back after the 30 seconds expired, but with such clumsiness, as if they had lost their flying skills; for, some were pirouetting in the air as if unable to decode their own echolocation signals; while others,—as if blinded by the returning sunlight or as if unable to detect the frequency of their own transmissions from the surrounding noise, like a radio receiver that's out of tune, or whose signals were being "scrambled,"—circled trees and mountain sides to then smash into them, as if attempting to enter into a cave that was not there.

Television stopped operating. On the radio, John heard that the "quake" or "jolt" occurred because there was a cessation or temporary interruption in solar gravitational activity on all the planets; as it were, only the sun's radiant emissions keep the solar system together. In short, the sun had stopped "shining" or operating for 30 seconds and its streams of rays, beams and particles, and binding energies, having ceased to function, caused a disruption in planetary dynamic equilibrium motions such as revolution and rotation, and hence, in gravity as well.

The gravitational field of energy exerted by the sun temporarily came to an end. There was no longer any "bond" or "binding energy" between the sun and the planets—for a few seconds, even depending on their respective mass, they were no longer "together," as a solar system. "Helter-skelter" they gyrated in deep space, "negotiating" an adjustment to the new radio-magnetic dynamics.

And on the Earth, at first, oceanic tidal wave continuum, atmospheric weather activity and landmass tectonics had undergone "a systemic slow down"; then, when the energy field provided by the sun returned, they engaged in "catch up" so to speak—hence, the "jolting, jerk-and-go effect."

Consequently, all Earth-system events and processes were suddenly "accelerated" to a maximum, pushing and pulling, attracting and repelling—oceans convulsed, their currents forming eddies with deep vortices, as the atmosphere groaned with violent whirlwinds that uprooted trees or "rushed them barren of leaves." The ocean-dominated landmass compressed the tectonic plates with such great force that promontories or rock formations from deep within the Earth surged through the surface to form new landscapes.

When the sun had restored its field of energy on the planets, this operation "shook" or "jolted" the Earth back into its regular patterns of motion and position—like slowing down while driving a car and restoring acceleration again as passengers feel "the jolt," but on a greater scale, the planetary frame of reference. Where the solar frame made contact with the Earth frame, there, were formed the most disruptive gravitational interchanges causing the Earth to undergo tremulous gyrations that registered in all its components and structures as violent cosmic quakes.

These velocity and mass differentials, along with magnetic field forces that bombarded the Earth, combined to impel the planet into a convoluted orbital path around the sun—with force within force, momentum within momentum, strain within strain, turbulence within turbulence, rising, building and climaxing—compacting the atmosphere, compressing the hydrosphere and contorting the landmass as the whole planet leaped forth in staccato movements to regain its spherical shape.

Sun spot activity was at its peak maximal climax. Scientists surmised in saying, "Perhaps the sun was experiencing disruptive turbulences of a kind we had little knowledge and which may have caused a type of short-circuiting in the 'continuum chains' of its nuclear reactions, like a humongous sun spot or nuclear storm system that engulfed the whole star—hence, the interruption in radiating activities and the 'jolt' experienced by the Earth afterwards. Perhaps it is comparable to when lightning strikes interrupt the flow of electrical current in households, and then flow is restored shortly thereafter as appliances are turned 'on' again and their functions return as expected."

And emergency preparedness agencies recommended that locally people engage in damage assessment and follow proper procedures for reporting and taking remedial actions; that every one should "brace for shock" and remain calm while keeping informed of developments.

"And until further analysis of solar activity is performed, no additional information can be given at this time."

That was the last sentence heard by John Trinklung from the radio announcement regarding the darkness that covered the whole Earth for about 30 seconds on that day in 2009—when only one longitudinal hemisphere was supposed to be in darkness, but not all the Earth. He remembered his son's comment about the sun, and kept thinking, "What could be happening within that 'big light bulb' that we should know about? 'Damage assessment?' 'It looks like the damage has been taking place for about a decade without us noticing. The sun—that 'big light bulb' in vacuum Space may be experiencing its 'last pangs of death'—a dying star! Will it 'burn out' or just 'fade away'?"

*   *   *

4

THE DISRUPTION OF SOLAR GRAVITATIONAL and radiant activity had dislocated the orbiting artificial satellites upon which television broadcasting so much depended; and it seemed that people in large urban areas who utilized "rabbit ears" or other types of antennae systems were able to obtain television programming. Television networks diligently endeavored to recover service with a few satellites after multiple attempts to re-calibrate their orbital path and re-orient their beaming signals were successfully performed.

Universities, "think tanks," schools, even families, and our government at all levels of emergency preparedness, readiness, response and delivery, were engaged in "solar dynamics thinking." What happened on that day, could it have been predicted as in "space weather forecasting?" Was it a "fluke" or "glitch" in solar plasma mechanics or did that event forebode a more ominous cataclysm? And how do we prepare for it when it affects the planet as a whole? Questions like those inspired great debates across the world as many people took opportunity to "brush up" on their knowledge of astrophysics.

Every one heard that a big tsunami had hit the coasts of Japan, Australia and New Zealand, due to plate tectonics displacement activity that reverberated into the hydrosphere—thus, oceans absorbing the larger portion of the shockwaves by re-translating them into the greatest tidal sea waves ever recorded in recent times. The devastation was so immense in terms of houses, buildings, roads, farms and equipment destroyed as well as lives lost, that it will take at least a decade before the affected areas fully recover.

Globally, and for good reasons, the greatest concerns have remained food and energy—the foundation for civilized human industrial and productive activities. At this juncture, national governments began to re-think their approach to problem solving in solar dynamics by redirecting their research activities towards more detailed solar activity observation and Earth-systems effects analysis.

At the United Nations, a conference addressing associated worldwide problems committed financial, material and personnel resources to delegating various research functions in accordance with available geographical facilities to which were assigned specific areas affected by the last solar conflagration.

Nations differed in economic development and resource utilization for comparative advantage, or rather for "opportunistic involvement" or "world engagement." Some nations excelled in advanced technology while others, having less technological development, however, possessed great manpower availability. They thought that geography should play a role in situating experimental apparati, materiel transportation and personnel deployment since this worldwide concern touched the lives of all human beings who shared the planet.

They reached the conclusion that explosive materials, chemical plants and nuclear facilities, and the like, should receive the greatest amount of security and surety in the prevention of accidents, loss, or catastrophic damage. However, it was unanimously agreed that local autonomy in sensitive response to self-government and conventional comparative advantage in allocating roles and analytical resources, should be kept in mind as they implemented the motto that "a global threat demands a global response." For, Nature and the sun could not be identified as the proverbial "enemy" though the threats were real as human lives and property hung in the balance.

That time however, the threat was really "extra-terrestrial," that is, it was "outside of Earth," from the sun itself—the very star that gives life-support systems to planet Earth as the only life-planet in the whole Universe.

Cultures all over the Earth and their forms of government began a redirection of their planning activities in whole new ways never encountered since before the very foundation of the world. For, changes brought about by disruptions in solar activity were systemic in nature; and a whole-energy system like the Earth, which presented especially complex sub-systems still under study at that particular point in time, was not even totally understood or known in all its multi-factorial contributing parameters. The sun had to be attended to; but more, the Earth had to be cared for simultaneously as the sun was being studied and researched and perhaps, even, "anticipated."

The majority of research facilities, including solar observatories, weather reporting organizations, space analytical agencies and university research centers, all, were contributing to the understanding of solar events and Earth effects in a way that conformed to the needs of agriculture, construction, travel, communications, medical care, maritime activities and commerce, and continuum industrial productivity in general, with public reporting conveyed with utmost care in prevention of panic, fear and rash action. The crucial questions they were asking were, "What will the sun do next?" "How can we prepare for it"?

On Wednesday evening, a consortium of concerned scientists, led by the astrophysicist Peter Barlotuk, gathered at the University of Illinois, Champaign-Urbana, in order to ponder the ultimate question of physics for our times—was it the sun's final death or "Terminal Entropy," so prominently configured in mass media and astronomy programs to which the public was accustomed? Or, was the sun dealing with some energy variances due to re-calibrations as necessitated by "Functional Operational Entropy"?

In other words, how was the sun's last "radio-magnetic activity" to be classified—was it like a "solar flare," or "solar wind," and thus, we had no need to worry about cataclysmic disasters? Or, was it something so totally out of the sun's ordinary operational re-adjustment repertoire as to confirm its path towards "Terminal Entropy," 'star-death' or 'sun-death'?

Their reasoning, prepared by Peter Barlotuk, took this form of logic: "And that was no eclipse—the sun's light was not blocked by the Moon, the Earth or any other space body! It did 'go out'—and that was a global event! Now, if the sun is indeed 'dying,' right before our eyes, in the now, in the present and not in some future so-called 5 billion years, then can we determine from its subsequent activities how long it will take,—given its mass, core-fuel dynamics, electro-plasma mechanics, convection processes, fuel quantity, strength of magnetic field etc . . . —before its imminent death overtakes us? And, if it is 'dying,' what can be done to avert the extinction of the Family of Man, the human species? We, the human family, we need a direct theory of entropy, a specific time-span and an appropriate constructive, saving resolution."

In Barlotuk's summary, it was decided that "in the short-run granaries should be filled, food storage facilities upgraded, fuel supply structures reconditioned to accom-

modate surplus accumulation; and emergency preparedness protocols and procedures revisited for effects-sensitive
responses addressing each respective locale that presents
its own kind of potential endangerment. For example, a
locale with an electrical generation nuclear power plant
should expect the benefit of a nuclear disaster response
team fully geared up through competent training, and com-
pletely equipped scientifically and medically, logistically
and materially."

They thought that it would be wise to have a set of reporting or communication protocols that began at the low-
est local level, with pertinent information going through the
county and state levels for appropriate response teams, up
to the federal level, depending on the complexities, severity
and intensity of predicted, and actual damage and loss.

"The Input-Process-Output principle ought to be
operating at each level, thus providing ample observation,
confirmation, data analysis and response options for each,
as the complexities of the situation would govern final disposition. But time is of the essence in damage control procedures. For, preparedness and time make for operational
readiness."

Peter Barlotuk, being first entreated with a permission
request, was approached by a young man with a piece of
paper in his hands. He summarily read it and addressed the
assembly in these words. "Friends, we have a new development. It is 8:30pm here in Illinois, that is U.S. Central Time,
and the sun has already set; however, it has been observed,
confirmed and reported that Aurora Borealis is now occurring, not just at the Poles, but as far down as the Equator.
This is not a new event in the sense that it has happened in
the past before, but given the current situation, it appears

that there is such excessive solar coronal mass emission activity as to go beyond the bounds of expected Earth-effects. No dangers from a simple 'light show' except that it might trigger changes in atmospheric layer processing of ionized gas penetration for cloud formations, the condensates of which are necessary for rain, replenishment of the water table, the planting season and soil fertility."

"Because of what occurred a few weeks ago, all processes, events or occurrences have to be driven to their logical conclusions; and the Earth can no longer be regarded as a so-called 'closed system.' The earthly Input-Process-Output mechanism has an intrinsic and inevitable connection to the solar Input-Process-Output mega mechanism. Every thing is important, and yet, not every thing can be factored in, with the same priority."

"And, in anticipation, a crucial determination has to be made regarding how long the reciprocal 'reaction chain' will hold on Earth, for purposive quantification and qualification of sub-system affectations, for example, in terms of predictions in ecological and atmospheric developments, hydrosphere tidal activity, tectonic renditions, and biosphere forecasts etc . . . within the periods of accentuated solar impacts and outputs that might disrupt Earth systems as a whole."

Paul Mirdewvell, a geo-chemist and bio-physicist, motioned for permission to speak and began thus, "We have unearthed some terrifying questions, ladies and gentlemen, the answers to which will determine the destiny of our lives in this world. We, however, are hoping these episodes of solar upheaval will not progress into its final demise. For, the last two instances of star turmoil countered each other as if torn by internal nuclear chain processing conflicts. First, there was no sun; but then, there was too much of it, in a manner of speaking. In the second instance, is not this in a way the exact opposite activity to what happened before, when the sun 'went out?' What I mean is that, it has been theorized that the previous event could have been caused by a giant solar storm system, like a giant sun spot, engulfing the whole solar complex and then collapsing its gravitational field so as to direct all radiant activity inwardly—and thus, no external emissions and the sun "going out;" whereas excessive coronal mass emissions like the great solar winds that caused Aurora Borealis tended to activate processes that are outwardly or externally directed rather than in-wardly imploded, perhaps heralding the augmentation of the sun's regular magnetic field strength. First, the sun emitted no light—as if it totally disappeared; but now, it's ejecting so much 'plasma rain' as to cause Aurora Borealis at both Poles."

"This 'oscillation' shows perhaps that the sun is return-ing to its regular core thermal dynamics and nuclear plasma convection activities, which will then restore stability, and hopefully, equilibrium to total-Earth systems."

Philip Karbidek, a physicist, who appeared to not have shared such optimistic assessments totally then added, "Such coronal mass ejections, however, have more or less been pegged to the eleven-year polarity change cycle which is sometimes accompanied by extreme sun spot activity. One would assume these activities have counter-effects be-cause they are opposite in operation—such as contraction and expansion, collapse and expulsion, of which convection is made, and thus, the absence of explosive spectral emis-sions in the first instance, and the presence of solar winds in the second."

"Plasma energy which powers stars like our sun consists of ionized gases produced at very high temperatures that contain about equal numbers of positive and negative charges; and plasma energy is an excellent conductor of electricity, hence, the sun's electrical dynamo. This dynamo is very powerful because it engages not only electricity but also radiation and electromagnetism. The atomic-ionized plasma fuel is a radio-active complex in which electron interchange is contrived by gravitational pressures and magnetic fields and thermo-nuclear temperatures."

"Solar energy as radiated ionized plasma, is first processed via thermonuclear fusion, and then 'cooled' into atomic energy—thus, the sun recycles its own atom-proton nuclei fuel. The sun is powered by variants of pressure and temperature just as our atmosphere (the whole Earth, rather) is powered by variants of pressure and temperature driven by electro-dynamics, which contribute to the formation of tornadoes—these processes thus creating atmospheric convections from which emerge hydro-electric catalysts for storms and wind systems."

"The sun engenders its own 'internal tornadoes'—ionized plasma and atomic energy are joined by nuclear convection, the upward and downward transfer of thermonuclear electromagnetic radiation,—as ionized plasma is fused and then 'cooled' into its component atomic energy structure. Solar thermonuclear convection is caused by electro-magnetic 'internal tornadoes' created by chain reactions fueled by the inter-transmutations between hot ionized plasma and 'cooled' atomic energy."

"Thus, the sun is a continuous 'raging tornado' devoured by such tremendous temperature and pressure, radiation and electro-magnetic momentum-force, and gravity differentials as to dwarf any conception we might develop regarding its internal operations."

"So how do we know if it is a re-normalization of regular activities, as opposed to a stop-gap process that will lead to another 'black out'? For if these are activities that impinge upon the sun's gravitational field binding-energies, then the processes from which these effects are emerging could be 'in toto systemic,' and therefore symptomatic of a 'kind of death' for which we will have little preparedness or response."

"The sun's gravitational field is not only essential to its own survival but also for the 'in-toto' binding energy that holds the whole solar system together. A collapsed gravitational field—caused by an all-encompassing sunspot—could have directed all radiant activity inwardly, thus 'over-cooling' the photosphere or solar surface, which then blocked all emitted radiation; but why temporarily, so that equilibrium is restored thereafter, at least from what we can observe and feel? And how often will that occur in the near future—or was it a one-time re-adjustment in solar core plasma and nuclear convection dynamics?"

"I mean, the laws of thermodynamics also apply to the sun's nuclear inferno, do they not? Is the sun running out of fuel? Is it suffering from periodic cooling so extraordinary that it affects its ability to retain nuclear chain reaction continuum for sustained radioactivity, heat energy, and gravitational projections for radio-magnetism? Or is it being inwardly 'overheated' so disproportionately as to need 'venting off' in order to remain as a star?"

"Of even greater concern, my friends, is that, if the sun's magnetic field is being maximally affected so as to impact Earth magnetic field electro-conductivity and gravity determinants, could not the Earth 'fall away from its orbit' and collide with another planet?"

"Or, what if next time the Earth collides with another planet that itself 'ran out of orbit'?"

Paul Mirdewvell, pausing for a few seconds, as if to let these horrifying questions "sink in," offered his own trepidations, "What if the Earth lost its ability to retain atmospheric temperature due loss of synchronized synergism between hydrosphere thermal absorption, core electro-conductivity and dipolar magnetic field radiation interception? After all, Aurora Borealis had reached the equatorial line separating the two hemispheres, 'merging' at the equator—as if the magnetosphere was 'weakening' its barrier against the onslaught of extremely intense solar radiation. Could that allow 'cold space' to overcome this barrier to engulf the Earth so as to engineer a 'cryo-state'? And therefore, could Nitrogen, 'a cryogenic liquid,' which constitutes more than 80 percent of Earth atmosphere, undergo 'cooling' so to speak, thereby neutralizing the thermal effects of less voluminous gases such as Oxygen, Carbon, and Hydrogen, that sustain livable temperatures? God be with us, with you, ladies and gentlemen!"

It was already 10:30pm and the meeting's questions hovered over every one's head, as if in anticipation of answers too frightful to entertain. The conference adjourned with promises of reconvening at various sites in many other States of the Union in order to spread dissemination of information in an angle as wide as possible given the opportunities of current telecommunications technology. But these local meetings were patterned after a model that was intended to survive even in the absence of satellite technologies, in case solar activities disrupted the grid of telecommunications on which the majority of organizations, businesses and government units had come to rely for transmitting, relaying, receiving and responding to information.

*   *   *

5

O N THURSDAY MORNING, John Trinklung and his wife
Claire were watching television when they heard what
happened to a truck driver in Nebraska. Kirk Nopherdon
was driving his semi-truck at 65 miles per hour on the highway as he was maneuvering to change lanes when suddenly
the whole thing lifted about 4 feet from the ground, crossed
over the embankment on the right side of the road to then
plunge 50 feet down and fall on a farming enterprise below. The gas tanks caught on fire from sparks flying from
contact with highway metal-works, and the truck exploded
as it slammed against the ground. Nopherdon was killed
instantly but no one else was injured. This report came from
drivers who witnessed the event first hand; "Our prayers are
with his family," they added. They questioned why this did
not happen to them because they were also on that same
highway.

The television network showed the broadcaster speaking to scientists who explained that "When the sun's magnetic field 'went out' it also affected the Earth magnetic field,
at least temporarily, so that gravity as pegged to center of
mass dynamics, in effect, was no longer G-1 on Earth, but
less—for though the Earth had continued to revolve around
the sun and rotate upon itself, as it had encountered nothing solid to stop its motion, it could not have continuum
acceleration until the sun 'returned again.' With gravity less
than G-1, the semi-truck became 'relatively lighter' than it
usually was. Do we not remember how the astronauts who
landed on the Moon 'walked' or appeared 'to float' or 'jump
slowly' on that day in AD 1969, because the Moon has only

about 17 percent of the Earth's gravity? Then, what percentage of Earth gravity was operating when that semi-truck was lifted up in the air, God only, knows."

"And thus, the semi-truck just happened to be traveling at the exact velocity with the correct momentum mass to be accelerated by an emergent G-force substitute that could elevate it from the surface of the ground to then catapult it through the air as if it could fly—that must have been a specific application of G-force reduction, just as tornadoes take place in a circumscribed volume of air within a specific surface area after which they 'fade away.' The shock wave received by the Earth from in-toto relative sun-Earth-Moon-planets dynamics, could have operated like a 'rubber-band effect.' For, the Earth is a complex whole-energy system with "hidden" or "not yet known" variables, factors and determinants that can cause many heretofore unknown natural forces, processes and events to emerge due to newly propagated solar gravitational dysfunction."

"If you stretch a rubber band to its maximum elasticity, holding its left end with your left hand and its right end with your right hand, and then suddenly release its right end from your right hand—that right end will rush to the left end to then strike your left hand and cause a 'shock wave.' So the semi-truck might have experienced an unusual lift from a shock wave occurring when the sun's field of gravitational energy returned to its fullest exponential power—after it had been absent for a while thus delaying normal G-1 effects. Witnesses on the scene reported that they thought they saw the trailer 'sticking up' in the air for a few seconds to then slowly lay on the ground, as if falling 'in slow motion.'"

The television network broadcaster then announced that funeral services have been arranged for Kirk Nopherdon by his surviving family, and affirming the knowledge that Americans are a generous people, that donations would be welcome to help his wife and children until they regained their financial stability.

The next news was that at a general hospital in Minneapolis, Minnesota, a doctor is said to have reported that one of his patients, a 51 year-old woman, Brenda Torlask, had no blood circulation for about 3 minutes during which the doctor quickly decided to put her on blood-dialysis machine and ventilator that would thus circulate her blood and provide her with Oxygen to breathe.

Blood circulation stopped while the heart was still pumping, which elevated her blood pressure but did not kill her, as the opposite effect took place right soon thereafter—the heart temporarily stopped pumping blood while circulation continued during these same minutes, thus, the whole organism alternating between pumping and circulating. Consequently, there were interruptions in both heart pumping and blood circulation, but within a 3-minute time span that allowed metabolic synergism to ensure continuum, while her body processed these alternations. Somehow, she lived through it all. An electro-physiologist suggested that she might have experienced a gravity-effect greater than Earth G-1 in order that slowed-down blood circulation might have simulated an interruption in circulation as the heart was pumping, which was then matched by restored circulation as the heart itself stopped its pumping muscular contractions. And because blood pressure was slightly elevated, and blood-Oxygen content seemingly not exhausted, body parts and organs continued to receive Oxygen for functional survival.

Apparently, the blood within her arteries, veins, vessels and capillaries was still rich in Oxygen content though the blood had stopped circulating—thus having the analogous effect of "breath-withholding," as if she had held her breath for a while after which she would resume breathing again. It did not affect her brain functions—neurotransmitters, electrolytes, and synaptic cleft activity appeared unaffected. The woman was kept overnight for further observation and treatment.

John Trinklung then turned to his farming newsletter in which the apparent drought or lack of precipitation was treated as a solar dynamics phenomenon that was unusual for seasonal patterns to which they were already accustomed. John and Claire noticed that though there was unbroken cloud cover and the sky was in continuous overcast, with gray-dark clouds, no rain fell. And this has been occurring for a couple of weeks now, with only "dry lightning" occasionally illuminating the skies.

In the newsletter, scientists surmised that, "For reasons still unknown, the Earth ionosphere has not retained electromagnetic radiation for processing towards lower atmospheric layers, bands or strata which are involved in electro-static 'energization' of cloud condensates for lightning, thunder and thus precipitation; that perhaps the 'wrong kinds' of clouds were being formed rather than 'rain clouds' or 'nimbus clouds' due to the extreme relative dryness of the atmosphere, and to extremely low hydrosphere evaporative cooperation."

"Variations in gravity away from G-1 will have had dysfunctional effects upon the Earth as a whole-energy system, dislocating and displacing the normal occurrence of all its processes—if cloud condensates or water vapors are relatively heavier when compressed against the ocean surface, they would not rise into the atmosphere to form clouds. That might have to do with the Earth magnetic field, core electro-conductivity and solar gravitational spectral emissions that are no longer in synchronized relationship for sustained regular gravity effects on the Earth, hence, affecting all major processes in the ecology, the hydrosphere, the geo-sphere, the atmosphere and the biosphere."

"All variations from Gravity-1 parameters will cause aberrations or deviations away from normal modus operandi; that, in turn, will have a deleterious impact upon all hydrosphere mechanisms that are evaporation-sensitive. Synergism for 'system energetics' between the sun and the Earth works best, with synchronization of all gravi-metric forces and electro-motive variables that factor into Earth life-support systems operating under optimum conditions for meteorological performance."

"When the sun's dynamic equilibrium is disrupted, it also has harmful repercussions for the Earth whose only star is the sun from which it obtains all necessary inputs for functional operational equilibrium at normal gravity, such as centri-vectored, curvilinear motions like revolution and rotation, magnetic field, hot iron molten magma core, core electro-conductivity for magnetic field sustenance, 'Gravity-1' pressure and temperature ranges, mass-specific motion-sensitive processes like the jet streams and El Nino etc . . ."

"Functional operational equilibrium at 'Gravity-1,' or normal gravity, is maintained by these two centri-vectored, curvilinear motions—revolution and rotation. Revolution around the sun is centripetal, meaning 'pulling' all things towards the center of the Earth; and rotation, is centrifugal, meaning 'pushing' all things away from the center of the

Earth. These two major counter-positional motion-forces of 'push and pull,' with momentum, magnitude, direction and velocity differentials, animate and permeate all Earth 'Gravity-1' processes, from atmospheric jet streams to oceanic hydrosphere currents, from rain falling directly or straight down towards the surface of the Earth, to an airplane being able to travel horizontally within the atmosphere with engine power that counters drag and opposes resistance to motion."

"Even the growth of a plant straight up into the air necessitates these two opposite forces so that the seed remains in the ground and the sprout will seek to 'move upwards' as it grows in search of light for photosynthetic processes via which chlorophyll, glucose molecules, and cellulose fiber, are manufactured."

("An example of curvilinear centrifugal force or rotational force is a phonograph or record player on which one can lay a 'plastic pebble' at or near a record's central hole or center of mass, to notice it 'fly off' the record as the velocity of the rotational vector will move it from the record's center or hole, towards the record's outer periphery; another force, the centripetal force, or revolution velocity vector, would be necessary to counter that centrifugal force in order that the 'plastic pebble' would remain on the record player as it would be redirected towards the record's center of rotating mass. Another example is the centrifuge with which astronauts are trained, which could be modified to have seated astronauts also rotate on their own seating axes in order to mimic two counter-vectored forces at the same time, to test how many g-forces away from 'Gravity-1' they could sustain without losing consciousness, and what kinds of movements and motions they could engage in at the same time. At present, they are just 'strapped into' the seat without being able to initiate any movement or motion. Adding the 'seated rotational movement' would have come closer to duplicating Earth-like G-force motions for training purposes and body-effects investigations.")

"On Earth, all things, objects, processes and events operate by the 'sensitive metrics' of gravity towards the center of the planet. If such 'centri-vectored,' gravity-sensitive metrics are disrupted, then human beings are liable to witness such uncommon or unusual phenomena as has been happening for a few weeks now."

"Therefore, functional operational equilibrium is sustained by Earth gravity and its dipolar magnetic field, the vectoring pressure-velocities and momentum motion-forces of which are maintained by continuum solar gravitational radio-magnetic binding energies for the different atmospheric layers, bands or strata to "process" solar radiation in order to statically electrify cloud formations which then will operate to produce lightning, thunder, and consequently rain."

John and Claire Trinklung had hoped that the writers would have attached, for future expectations, a sense of certainty to their explanatory declarations. However, the Trinklungs also knew that many internal sun processes were inaccessible even to the best, state of the art technology, the limitations of which might forestall sure knowledge of cause-and-effect relationships that effect disruptive changes in solar activity, and thus, on the Earth.

*   *   *

6

THE TRINKLUNGS HAD BEGUN to farm about a decade ago but they have never seen such a phenomenon, even on the farm. John was speaking to Claire his wife, on the porch, when Marc and Jeanie, Marc's younger sister, joined them, but with an excitement seldom demonstrated before. It was about Marc's school experiment, actually with left-over meat from dinner eaten two evenings ago. The school did not really assign it to him; but given what has been going on with the sun and the Earth for the past two weeks, teachers had encouraged students to report any unusual happenings while refraining from inciting make-belief or forged events to gain attention.

"Dad, Mom, you should see this. It's amazing. You remember the meat left-over from the day before yesterday—from two days ago. Today is Thursday; I left it in a covered bowl on Tuesday evening in the shed in order to see what bacteria and parasites would do to it. It's 7:00pm now; so that's about 50 hours ago because we ate at about 5:00pm on Tuesday. But nothing happened to the meat. I mean, for two whole days—it smells as fresh as when I put it into the bowl—it's as if decay had stopped, or as if bacteria could not 'do their job,' which is, devouring the meat to the point where it would start rotting and smelling. What do you think, Dad, Mom? Do you think that's significant, that something 'unusual' or 'strange' is happening?"

"Can we see it?" asked John.

"Yes!" entreated Claire. That's a good idea. Then after smelling and evaluating it, we might have a better opinion on the subject."

"Ok, Mom, Dad, I'll show it to you."

John and Claire followed Marc and Jeanie to the shed—Jeanie was 14 years old; she entered high school this very year. All came into the shed as Marc turned the light on for better navigation. John and his wife stood side by side as they reached for the covered plastic bowl to which Marc pointed on the shed's table. Claire proceeded to open it very carefully as John watched in earnest. They looked at it and smelled it very closely. Picking up a fork from the shed's table, Claire lifted the morsel of meat very slowly to look underneath it, as John held the bowl in his two palmed hands.

"Did you add anything at all to the meat before you stored it?"

"No, Dad. I did not. I mean, it's a piece of steak. I stored it the way it left the house on Tuesday."

"What do you think, honey?" John said, talking to Claire.

"Well, this meat is as fresh as when we ate it, at least from what I can tell and smell. Wouldn't you say so too, John?"

"Yes, it's my opinion too. But we are not scientists in the true sense of the word and have no laboratory equipment here. I wonder if we could call the school or the University of Illinois, since your teachers had told you to report anything 'unusual' and then they could examine it in their labs."

"Yes, Mom, Dad, let's do that, please. I want to know why after more than 48 hours the meat has not degraded, decayed, or rotted. I am doing well in biology class and that could count towards by class project if Mr. Pirkalmehr agrees."

"And, it's on-going," added John. "No smell still."

Marc's father, John, resealed the plastic bowl wherein the piece of steak was stored, and turned the shed light off.

They summarily returned to the farm house as Claire was engaged in retrieving the school's phone numbers. But it was nearly 8:00pm now, on a Thursday. They decided to call Marc's biology teacher at his home; he had connections with the university and if anyone could do anything, he would certainly be a good candidate.

Luke Pirkalmehr had been teaching biology for a few years and took care to listen to the Trinklung's report, after which, he also talked to Marc, and to even Jeanie who was also a living witness. John Trinklung said a few words about his son's assiduous and diligent adherence to his teachers' instructions, for which Luke Pirkalmehr expressed appreciation too.

"Is there any way you could bring the sealed covered bowl with the meat to the high school to me and then I could talk to other scientists there for a laboratory examination; hopefully that would tell us something about why those little microbes were not or rather are still not hungry, so to speak."

"Just a minute, Luke, please."

After a few seconds talking to his family and his wife, they all agreed that would work out extremely well, especially for Marc who made the discovery and who was so excited about the prospects of being engaged in an important microbial experiment.

"Luke," said John Trinklung, "We would be glad to do that. We can meet you at the high school in about 40 minutes. And we will bring every thing. Is that ok with you?"

"Fine, of course, John. I'll see you in a few minutes then. Thank you."

"O, we thank you, Luke," enjoined Claire. "And we're so glad to be helping Marc out. He is so excited about this and we too, in a way, for we appreciate his interest in biology and other subjects, which will 'serve him well' if he decides to go to college. Thanks again, Luke, please forgive our interruption of your evening."

"No, you did not disturb us. We, the whole family, were just catching up on today's events anyway. So that completes the circle in a way," said Luke.

"Thank you so much," replied Claire.

"By the way, John," continued Luke, "I just thought about something in passing. Have you had any kind of weather or atmospheric activity in the area at all for the past three days?"

"Well, we have a lightning pole at the south end of the property because of "dry lightning" that is common to this part of the State. I mean, there is lightning but no rain," said John.

"How far is that lightning pole from the shed, John?" asked Luke.

"O, Luke, not that far, about 50 feet, but far enough that no fire has ever taken place on the property. And that pole seems to be doing its job well," replied John.

"One more question, John, would you say that you've had uncommon lightning activity for the past three days that you can remember?" inquired Luke.

"Well, Luke, it happens so often that we don't pay any special attention to it. But yes, I would say that I noticed a slight increase in the frequency of lightning, even during the past two weeks or so," summarized John.

"Thanks, John. We will let you, and the whole family, know about the lab results. I find this very interesting too, to say the least—a nice break in the teaching routine! We'll talk again soon. Bye," said Luke.

"Thank you, Luke. Best regards to your family. We'll look forward to it. Bye now," replied John.

The whole Trinklung family drove to the high school where Marc met his biology teacher, Luke Pirkalmehr, to deliver the closed, sealed, covered plastic bowl and its precious cargo—the morsel of non-rotting meat, with expectations too great to describe and plans too noble to explain.

*   *   *

7

"THIS IS AMAZING!" THOUGHT Burk Marnvat, a passenger with a degree in The Philosophy of Science, as he gazed into the clouds passing by his glass window.

"Here we are in a machine, flying within a ball, made up of rocks, water and gas, 'suspended in space'; the sphere is moving around the sun and so are we while in it.

We're 'moving' in a plane; the Earth is 'moving' through space. The plane is moving in Earth 'air space'; the Earth is moving in 'air-less' space! We're all moving in space—the only difference being our atmosphere with gravity! Our space has air in it for us to breathe, and gravity to move with. Wow! Thus our own space,—the space within the plane, and earth atmosphere-space,—is also moving through 'air-less' vacuum space."

"The sphere is also rotating on its own axis; so are we, upon ourselves—upon our own axes! The sphere is revolving around the sun—so are we!"

"The whole Earth: We're moving within its own motions; we're in outer-space within its own space—motion moving within motion; space moving within space. Mass within mass—atom within molecule, molecule within organ, organ within a greater part or member, a part or member within still a greater body; how amazing this is!"

"The atom is within the planet, the planet within the solar system, the solar system within the galaxy! And, we are within all this—consciously examining it!"

"So mass-effects are within Mass-effects, gravity effects within gravity effects, motion within Motion, space within Space—Wow! That's the way Mass 'curves' or 'bends' Space!

I remember Galileo who said 'Every thing is made up of every thing.' So now, I could say, 'Every thing is within every thing!' Earth's geo-atmospheric dynamo is also generated via this process—an electro-conductive atmosphere—converting mechanical energy or 'motion energetics' into electrical energy!"

"O! How about the Sun! Radiation and electricity in super-system-synergy! Radiation, electricity, magnetism and gravity—what awesome radio-metrics! Pressure and temperature are only co-determinant controls! The sun's nuclear plasma molecular friction with humongous gravi-magnetic field-motion pressure-forces—as an electric super-dynamo! The mechanics of its magnetic radio-electrical 'synergetics' generate trillions and trillions of volts! What a power station that is!"

"If only we could understand all the linkages between solar plasma 'synergetics' and Earth atmospheric 'energetics'!"

"We are actually in an aluminum tube flying within a great ball around the solar system. We are suspended in Space. Why is it so, no one really has an explanation; some things we just accept, like a little seed growing into a big tree after it's been planted."

"Like one flying object within a greater flying object—plane within planet, air space within 'air-less' deep Space! And all of this is possible due to Gravity—an invisible force devoid of all substance!"

"A machine in a planet becomes a planet with a machine around the sun—traveling in a great trajectory that has the appearance of a straight line but whose path is really curvi-linear—together we're making a great ellipse around the sun while the Earth is making a circle on itself! And that's the Earth within the solar system—things within things moving in spaces within spaces! That makes for the greatest wonders in human existence. Wonder of Wonders! Only God could have conceived it, only God could have 'birthed it', only God could have created it."

The 727-Boeing jet was traveling at normal cruising speed at about 1:00pm on that Friday; it was approaching the city of Chicago in a flight path from Montreal, Canada, as the pilot's flight plan will have brought him to land at O'Hare Airport. Earlier that morning, "solar weather" reporting hinted at the possibility of coronal mass emission turbulences that might affect telecommunication satellites, guidance and control systems, radio and television broadcasting, and cellular telephonic operations.

However, no time period or duration was given for those activities. No forecasts were made concerning any particular locality to be targeted but that all precautions should be taken so as to minimize damage or loss even while regular commercial and other activities were ongoing.

The flight captain announced that the plane will have been landing in 10 minutes at O'Hare Airport. At about 8 minutes till landing, the captain assumed control of piloting, throttling up to re-adjust the aircraft's position, while the flight engineer was preparing for systemic checks pertinent to proper functioning of landing operations.

But the captain observed that, though the plane was no longer flying on automatic pilot, he himself had no control of the aircraft; the flaps and the ailerons were locked into one position—keeping the aircraft parallel to the line of horizon. It was as if the controls were all useless and that the plane was "flying itself."

"What in the name of . . . What is this?" muttered the captain.

They immediately informed the control tower at Chicago O'Hare Airport of these problems, including the probability of not being able to land. Given that the captain and the assistant pilot knew the plane had extra fuel, the captain did not proceed to reduce speed and engage in landing procedures.

"Unable to land," broadcast the captain to the O'Hare control tower. "Will try St. Louis, Missouri. Have enough fuel. Thanks."

"How about the passengers?" asked the flight engineer.

"An announcement is forthcoming," replied the captain.

Passengers were informed that the plane encountered some difficulties with electronic guidance systems and landing mechanisms over Chicago and that they were headed to St. Louis, Missouri, from which, after landing there, arrangements would be made to return them to their original destination in Chicago; that every one should remain calm, that every thing was under control.

"How about Springfield, Illinois," asked the flight engineer; "it's closer than St. Louis. However, the timing might be inappropriate for preparedness and readiness operations."

"True," replied the captain, "but, in addition, the question is, do they have the wherewithal, facilities and technologies to attend to our need in case of a forced, or even, explosive, landing? By the time we get to St. Louis, they'll be ready there. Furthermore, personnel in St. Louis have already been informed and are making preparations for all eventualities, including an explosion."

"What could have caused this captain? I am sure you also know, every thing checked out 'super' when we left

Montreal," added the flight engineer, who was also the assistant pilot.

"And even during flight, every thing checked out ok," replied the captain.

While the plane was on its way from Chicago to St. Louis, all personnel and passengers were in their respective seats, with their seatbelts on. Many were praying with their eyes closed; others were chatting regarding their last wills. But there was no panic or loud exclamations. Then, something else, outside of themselves, got their attention.

A passenger noticed that things began to lift or soar upwards from their initial positions in the plane. Every thing that was not tied down began to "float" so to speak, as if there was a "vacuum" in the aircraft.

One woman asked, "If there is a 'vacuum' in here, how come we can still breathe? I thought outer vacuum space had no atmosphere?"

One man retorted, "It's a simulated vacuum. Not the real thing. But we're not free falling; there should not be a vacuum in here. That's the way they train astronauts, by free falling in flight in order to simulate micro-gravity. But the aircraft is traveling normally at regular velocity. Why then are these things in the air around us? I hope this is not some kind of 'experiment.'"

"O, I don't think so," said another woman. "We're flying through the air. I'm sure there is some scientific explanation for all this. After all, this is an aircraft, the most sophisticated of all machines created by man. At least, we've not been hit by lightning!"

"You sound like you travel a lot," said another man.

The captain knew he'd better make another announcement: "Please remain seated with your seatbelts on. Some things are happening in this aircraft for which we do not yet have immediate explanations, but we are sure that the plane is in good condition and is air worthy for landing in St. Louis, Missouri."

"Try doing something to unlock these flaps and ailerons," the captain said to the flight engineer. "Whatever is holding them in horizontal position, we don't know. But they might get free by moving the controls in different directions as in preparation for take off or landing. We are high enough in altitude to anticipate maneuvers necessary to regain horizontal flight."

"Captain," said the flight engineer, "nothing is working to change their position. What could be holding them like that, we can't assess from the cabin. Perhaps I should go below and try to locate the cables that activate flaps and ailerons."

"Yes, go ahead. It seems that we have no other option," replied the captain.

Suddenly, as the flight engineer was motioning to leave, he was "jolted and bolted" back to his seat again as the aircraft gained lift to rise about 50 feet up, to then assume horizontality with flaps and ailerons unlocking in the process.

"Captain," exclaimed the flight engineer, "What happened? What did you do? Fifty feet up in the air! How did we regain the horizon line so quickly? We have to let them know in case there is another aircraft in this flight path?"

"I did nothing. It just happened. Thank God there wasn't. Thank providence you were the only one who tried to move. Thank God, for every thing is free now—the flaps and the ailerons. Let us return to approved altitude, that is, 50 feet below where we are now."

With control surfaces returning to pilot's commands, the captain immediately executed steps necessary to return the airplane 50 feet below their current position, in accordance with flight plan.

"But how can we explain all these events happening all by themselves, captain?"

"We'll, or, they'll figure it out somehow. But the main thing is that things have returned to normal. We are safe and can land in St. Louis without any further problems. Take a look at the passenger compartment and see how things are while I make an announcement, after which I'll talk to St. Louis, so that they can inform Chicago."

The flight engineer could not believe his eyes—the apparent chaos with baggage, cups, bottles, pillows, blankets, magazines, masks, spoons, etc. . . , everywhere, but with every passenger safely in his or her seat—a great relief. Stewardesses joined him, attending to passengers's needs and comforting those who needed relief.

"Excuse me," said a passenger, "How come we went up so fast. I got hit by a pillow on the head as these floating things were falling down."

"A pillow, thank God it was not something hard or solid that could have given you a concussion, sir. How are you feeling otherwise?"

"O, ok! But this lady here was vomiting a few minutes ago."

"How are you feeling now, Miss? Would you like something to drink or eat?" asked a stewardess.

"Something salty, please, with a cup of water. Thank you," she replied.

*   *   *

8

MATTHEW LANTERNOUGH WAS A semi-retired micro-biologist with a minor in geo-physics, a good friend of Luke Pirkalmehr, the high school biology teacher. Luke explained every thing to his friend to the last detail and requested his assistance in examining the piece of steak obtained from his student, Marc Trinklung. Since the high school laboratories had rather limited resources, they decided to meet at the university instead.

Luke Pirkalmehr obtained the meat from Marc Trinklung on Thursday evening and it was on Saturday morning at the earliest that Luke Pirkalmehr could receive any laboratory assistance from his friend. Thus, the meat would have been in that "suspended unspoiled state" for more than 72 hours since Marc Trinklung had first stored it in that closed, sealed, plastic bowl.

"Hi Matt, how are you doing?" said Luke upon seeing his friend.

"Great, Luke! It's so good to see you too!" answered Matt.

"Is anyone else coming, Matt?" asked Luke.

"Yes, glad of you to ask, because Heather and Robert are also coming," replied Matt.

"Excellent, I am looking forward to seeing them as well. It's been a while; we haven't gotten together since our college days," responded Luke.

Heather Sunkeltone was a microbiologist with a minor in chemistry and Robert Tillkamore was a bio-physiologist with an interest in astrophysics.

"Here they are," said Matt as he looked over unto the parking lot near the building where they were meeting.

"Wonderful," replied Luke. "We'll have an early start then."

In a few minutes time, they were joined by Heather and Robert who were also effusive in recognizing their old friends. After exchanging greetings and reminiscing on good old times, they decided to get down to business—the laboratory. Luke unpacked the plastic bowl with the closed lid wherein rested the piece of steak slated for examination.

As Matt opened the sealed plastic bowl, he thought he heard air "rushing out," as if there was a "vacuum" within the bowl. But the sound was so faint as to have almost escaped his notice. But the others confirmed to have heard it also, which may factor into their assessment of variables responsible for the steak's unspoiled status.

The laboratory was well equipped with highly technical instruments as well as a couple of electron microscopes. They decided not to cut the meat or sever any portion thereof. They thought it would be better if it remained whole in order not to alter any "initial conditions" upon which the meat must have depended to assume its state of suspended spoilage.

Operations began at 8:30am that Saturday morning, each taking turns at examining the meat and analyzing their findings. And it was not until 1:45pm that finally each and every one of them could have something conclusive to report, at least tentatively.

Matt began by saying, "This is absolutely, remotely impossible—I mean, what I am trying to reconcile with here. Do you know that the meat apparently has not spoiled, degraded, decayed or rotted at all, but that at the same time it is 'dead'? I don't even understand how to word this—it's not 'dead' in the sense that it has lost its suppleness or moisture; it's not even hard. But for all practical purposes, it's 'dead.'"

"Yes, I understand what you are saying," said Heather. "Would you say that it no longer conducts electricity?"

"Correct," exclaimed Matt. "No electro-conductivity! But how can this be?"

"Like the molecules had lost their capacity to bond at their outer electron orbits or levels, as if losing electro-static sensitivity or receptivity to energy," enjoined Robert.

"That's exactly the way I feel things are," said Luke. "I can't explain my understanding in your own words, but at the same time, I cannot object to any of your findings. The meat is not spoiled but at the same time it's 'dead.' It has no smell and has not hardened. What could have caused such momentous transformations in biological cells that had undergone regular cooking?"

Matt tried to figure out how to ask questions that would ferret out some logical scientific connections with possible "initial conditions" and the data presently at hand in the hope of formulating some hypothesis regarding this state of affairs.

"Would you say the cells have atrophied or ossified, or have undergone some process or other that would have led to the meat's suspended unspoiled status and non-electro conductivity?" he questioned.

"Matt," replied Heather, "The resistance to spoilage might be connected to non-conductivity since molecular bonding occurs due to electro-static charges between atomic electrons involved in a reaction between elements. These charges attract and repel to form the proton-proton, proton-electron and neutron-electron relationships complex. But even if we did agree with your analysis, what cause-and-effect relationships could we identify or establish that would explain the morphological processes?"

Robert said, "Matt told me about the Trinklung's shed, Luke, the lightning pole, 'dry lightning' and the 50-feet distance between the pole and the shed, etc. . . , and since we have talked about loss of electro-conductivity, then is there any reason why we should not bring in these factors which may not be extraneous at all, but conducive to better understanding the 'initial conditions' that might have caused these 'bio-morphic' cellular effects? Are you familiar with the Trinklung's property, Luke?"

"Yes, I believe so, Robert. I visited them a couple of times and do remember the overall arrangement of buildings etc . . . And it is possible that materials, things and objects in that shed might be sensitive or responsive to a field projected from the lightning rod, if any was created by the lightning bolts hitting the rod," replied Luke.

"Well," enjoined Matt, "an electric field will generate a magnetic field; and a magnetic field will generate an electric field; or, we can say, they are inseparable and that they go together; where one exists, so will the other, so to speak. Could these entanglements have caused the 'morph-organic' cellular transformations?"

"Possibly," said Heather. "So we have the meat in the closed plastic bowl on a table in the shed, a lightning rod generating an electro-magnetic field about 50 feet away from it, but that has a field of influence with a radius long enough to reach the shed, and a shed wherein are stored multiple things and objects made up of all kinds of materials, including fluids and metals. Is my recapitulation valid?"

"Of course," said Matt, "Then how would the electro-magnetic field work within the shed and upon its materials to affect the meat morsel in the manner as we have observed while the bowl remained untouched? It could have been a "field event" for if electrical charges were involved, the plastic bowl would have demonstrated some indications that a form of contact with the discharges took place. Did any of you find any bacterium or parasite near or on the meat?"

"I did," replied Heather.

"So did I," said Luke.

"How were they in appearance, texture and structure?" asked Robert. He added, "I did see a few also, but not too many. It's as if they were trying to avoid the meat but could not resist the 'fatal attraction,' so to speak."

Heather replied, "The bacteria I saw 'looked sick,' as if they had regurgitated after having lost motility to move, climb on, and 'attack' the meat."

"I would concur with you," said Luke. "But more, it's as if 'things were oozing' out of them; I have never encountered this before in looking at parasites under a microscope."

"What could that 'oozing' be," inquired Matt? "Do you have any concept, insight, idea or inkling, Luke?" he asked.

"Well," said Luke, as you all know, bacteria have walls which are destroyed by enzymatic action. Penicillin, for example, is extracted from molds that are antibiotic in operation. Could we be looking for something like that? But, how? What could cause an antibiotic effect in the meat, to ward off bacteria at the same time preventing the meat from spoiling—if indeed that is what's happening?"

Heather then added, "Well, how could the sun literally 'shut down' and 'go out' and then 'return' as if nothing had happened? How about Aurora Borealis at the Equator? These events did happen. Things have gone awry in this universe, the complete causes of which we have yet to fathom. But it looks like, that in order to have a determination regarding these questions, we'll have to cut into the meat anyway, whether we like it or not. What do you think folks?"

"Let's eat something," proposed Matt. "Then afterwards, we'll figure out how to proceed."

"Good idea," said Robert. "I'll order some pizza and we can eat in that room next to the lab."

Every one agreed and they took a well-needed break from the laboratory to enjoy a few minutes of rest while replenishing their energies with food, beverage and good memories.

*   *   *

9

THE CONFERENCE HAD OPENED early morning that Friday in the presence of National Transportation Safety Board staff, but with airline-corporation personnel presiding, as other commercial and community interests were also represented for as much fluency of information as possible.

The captain of the Boeing 727 was James Nytolfitz, and his assistant pilot or flight engineer, Jack Ralbiehl. They were both present with all stewardesses as well as some passengers who could make it to the meeting. All desired to understand the events that took place on the flight from Montreal to Chicago that was diverted to St. Louis, Missouri, due to reasons still unknown. For these causes, and solar dynamics effects on the planet, present were astrophysicists, geo-physicists, aerospace engineers, physio-chemists and materials engineers, crafts-integration specialists, among whom participated other scientists associated with the behavior of aircraft in atmospheric conditions of disruption, all, on hand, to assist in developing theoretical and practical explanations for the puzzled traveling public and all parties concerned.

Testimonies were taken from the captain, James Nytolfitz, the assistant pilot, Jack Ralbiehl, several stewardesses and many passengers, all of whom taking care to report all facts and observations so as to provide the greatest latitude to flight investigators and scientists for the prevention of air disasters and other spatially related damage control purposes.

Lucia Brundfield and Jeanette Carquonold, astro-physicists with experience in solar theoretics were consulted to provide their interpretation of the facts and observations, assisted by Lernell Brindock, a geo-physicist and Hernbank Vultig, a degreed professor in Earth sciences, and joined by a materials physicist and physiologist, William Pelville, and Dennis Kribonzak, an aerospace engineer and medical doctor. They all held doctorates in their respective fields of endeavor.

Lucia Brundfield began addressing the conference attendees by stating her credentials and qualifications, as well as introducing other scientists and their qualifications and credentials, including the two pilots of the Boeing 727, in order to save time, and so that, when it is their turn to speak, the attendees would have had been familiar with the biographies of scientists and engineers for identification and reference purposes.

Lucia Brundfield proceeded thus, "These events that took place on the Boeing 727 flight from Montreal to Chicago and then to St. Louis, ladies and gentlemen, present perplexing observations to say the least—for example, pillows, books, blankets, cups etc . . . 'floating' in mid-air as passengers themselves are fastened to their seats by seatbelt restraints, and the plane suddenly rising or being lifted with such thrust and acceleration as to propel it 50 feet upwards from its original flight path—all, without action from the captain or the assistant pilot—with ailerons and flaps 'locked' into horizontal position, and then suddenly released without warning."

"These are remarkable happenings or exceptional events whose cause and effect relationships tend to evade analysis, especially since these activities cannot be amenable to actions taken by anyone onboard the aircraft itself."

"As you all know, there have been many 'strange' or not-yet-known solar system events and processes, the dynamics

of which, we are still trying to learn and understand in order that we, as the Family of Man, the human family,—or as
some say, the human species,—might continue to live and
prosper on this planet. For example, the sun apparently 'lost
all its light' on that day when every where it was as dark as
night, though the time was only 2:00pm in the afternoon.
Another time, Aurora Borealis occurred all the way down
to the Equator, 'merging' into that zone, when it is usually
associated with the North and South Poles."

"Scientists have even discovered a wide range of varia-
tion in the strength of the Earth magnetic field—sometimes
it has increased, at other times, it's the opposite, it has decreased. In addition, the Earth gravity has also been found
to vary to intensities less than normal or less than G-1, and
at other times, to magnitudes greater than G-1, thus, causing
human beings, animals, and things, and objects on Earth to
experience 'gravity force substitutes' that might have ramifications for the health of human beings, the operation of machinery, the proper functioning of animal existence, as well
as for geological, ecological, and bio-sphere operations."

She then motioned to the other scientists to see if they
had anything to add. Dennis Krybonzak enjoined, "In accordance to the foregoing information, I would like to
submit to you that the Boeing 727 flight had experienced
gravi-metric and magnetically driven effects, the causes of
which are directly traceable to solar activity. It appears that
the sun is 'oscillating back and forth' between Qualitative
Conservation and Functional Operational Entropy—not
yet experiencing Terminal Entropy or final, total 'death.'"

William Pelville continued, "I would agree that these
events proceeded from solar gravity-sensitive processes that
affected the binding energies that keep the Earth magnetic

field and G-1 Input-Process-Output mechanisms in efficient operation. And it could be in that manner, or due to these complex effects, that the plane's motion was re-directed as a gradual, but vertical fall, from its original flight path, so that in appearance and sensation, that is, relatively speaking, a temporary internal 'vacuum' could be experienced by materials and things inside it, while at the same time possessing a breathable atmosphere for people to live by."

Janette Carquonold then posited, "Yes, a 'gravity-effect or force substitute' therefore acted on the plane to cause its motion to change in a vertical descent while maintaining equilibrium parameters that could cause discomfort if disrupted, however, which were not. It is perhaps that descent,—since all of you were in your seats with your seat belts on and your body could not feel or sense the vectoring change in motion direction—which engendered the simulated "vacuum" in the aircraft. And that is why passengers saw things 'lifted up,' 'rising' or 'floating' as they sat restrained in their seats. For it has been reported that the lady who was nauseous experienced vomiting after the plane was thrust upwards 50 feet, which all of you felt, experienced and sensed, and not during the witnessing of the 'vacuum effect.'"

"Indeed," added captain James Nytolfitz, "My assistant was just on the brink of moving when that lift upwards occurred. Well, the aircraft must have reached dynamic equilibrium at line of horizon right after that vertical downdraft took it down; for it re-stabilized right away but with flaps and ailerons locked into unchangeable position. I had not yet reached cruising altitude when the downdraft took us down in order to simulate micro-gravity or vacuum space. But the shock suffered by the plane or wind vortices experienced by the aircraft control surfaces while within these motion-forces, might have explained the temporary locking of these mechanisms which were restored upon the boost upwards."

"However, these haunting questions still remain: What could have caused these downdrafts at first and then a boost in the opposite direction—but up 50 feet above our regular flight path?"

"Since the sun appears to be 'oscillating' between Conservation and Entropy as Dennis Krybonzak pointed out," enjoined Jack Ralbiehl, the assistant pilot, "and as we have been apprized regarding the Earth magnetic field and gravity equivalents, then it is possible that a 'vacuum' or 'hole' was created by the mass and velocity of the Boeing 727 moving and accelerating into the atmosphere during disruptive emissions by the sun and associative reactions by the Earth, in Earth attempts to recover normalcy and dynamic equilibrium from solar storm systems. This 'vortex' or 'gravity-induced hole' into which the downdraft took the aircraft vertically yet without disrupting aircraft lateral stability might be analogous to a 'suction' or 'vacuum vortex' created by pressure, wind, and temperature differentials in raging tornadoes as they gather strength so as to accumulate into a circumscribed geo-gravi-metric tempest."

Janette Carquonold responded, "The vortex or 'gravity-induced hole' as you refer to it, into which the plane descended, assuming that is the only way to generate or simulate a 'vacuum' in Earth atmosphere, could have been engendered by temperature and pressure differentials that apparently had not registered with bio-metric sensitivities of anyone onboard, since every one agreed to have felt the upwards lift rather than the downdraft descent. The verticality of the downdraft is even a greater mystery—taking into account the mass differentials in that aircraft, it did not 'nose-dive,' so to speak. The force of the downdraft momentum on control surfaces might have been accountable for locking the flaps and ailerons mechanisms, and thus sustaining horizontality. And since the plane continued its motion as if on automatic pilot, then no one experienced damaged health or sensation discomfort. But when the 'vortex effect' ended, then something else was triggered."

Lernell Brindock joined in, "Yes, when that 'vortex effect' had ended, releasing the plane in full force, a momentum pressure-force greater than G-1 gravity force was created. Or, perhaps, in terms of Relativity effects equivalency, the plane's own movements may have left that 'vortex' or 'hole' below it with such great thrust, that 'it freed itself,' so to speak, rather than the vortex itself ending—the effect of course being similar, in that the plane was propelled upwards 50 feet. That could be a possibility since nothing was changed as the plane was still accelerating towards cruising altitude, including the aircraft's velocity, except that the flaps and ailerons were locked into horizontality. It's as if a rocket booster was added underneath the aircraft to lift it upwards in so fast a time as to startle the mind and make the imagination wonder. I mean, a Boeing 727 is a pretty heavy plane, and to have it catapulted 50 feet upwards into the atmosphere in such a short time is phenomenal, to say the least. I am sure they could use an engine like that at Boeing, if they could duplicate it."

Hernbank Vultig entreated, "Possibly yes, the 'vortex effect' contributing also to the aircraft's moving momentum, exponentially adding to its velocity. And this is during that uplift, after it was propelled upwards or 'broke free,' that things that were 'floating' began tumbling down—as in the case of a pillow hitting a passenger on the head. It could be compared also to a 'rubber band effect'—stretching a rubber band while holding one end and releasing the other end, or compressing a spring to then release or dilate it. It is not surprising that some passengers got nauseous then—during or after the updraft. Their bodies must have experienced a G-force substitute greater than G:1 or normal gravity, which of course, could have caused nausea, vomiting, cold sweats, and other discomforts etc . . . , because it affected electrolytic balance as well as neurotransmitter effectiveness in 'carrying' signals received from brain cells."

"And then, as the aircraft had stabilized at 50 feet upwards from its regular flight path, the captain said he noticed that the flaps and ailerons were functioning properly," added Jack Rabiehl. "Having been freed from the 'vortex effect' vertically, the aircraft's control surfaces must have unlocked and re-stabilized from the downdraft's 'drag' or 'downward pressure' which exerted counter-forces upon them while they remained horizontal—because the plane had received no other command or instruction from the captain. At such velocity and vortex-induced momentum, various forces of drag and lift were acting upon the aircraft, countering 'yaw' and 'roll.' Flaps and ailerons then did not go down due to these counter-forces, because the aircraft had ceased to climb, albeit vertically, stopping at 50 feet upwards from its previous position. I thought the captain himself had managed to unlock the flaps and ailerons, but got even a greater shock when he told me he had done nothing of the sort—that 'it just happened.'"

Lernell Brindock enjoined, "Thus, as the aircraft was catapulted 50 feet up in the air, while the flaps and ailerons were locked, horizontality was sustained due to cessation of upward 'drag' upon control surfaces, the sudden shock of which freed flaps and ailerons at the same time, with the captain taking control immediately in order to stabilize the aircraft."

Lucia Brundfield thought it was time to recapitulate what they had learned from this experience and concluded in these words, "Well, it seems that we have agreed that these events could have been caused by solar activity dynamics in conjunction with Earth response to equilibrium disruptive effects, which engendered gravity and magnetic field disturbances, akin to those that create 'tornado vortices,' or other Earth storm systems, for example, due to wind, velocity, acceleration, mass, temperature, momentum, and pressure, differentials. Though these are not totally satisfactory explanations, we will have to comfort ourselves with such preliminary evaluations until more concrete determinations can be made."

"Does any one have anything else to add? Our addresses, phone numbers, websites, and e-mail addresses are posted here for public comment and additional reporting of observations. Please do not hesitate to avail yourselves of these opportunities to advance our knowledge and understanding of Sun-Earth relationships which are so vital to human lives on this planet—especially for air flight and aerospace. Thank you for coming. God bless you all."

*   *   *

10

MATT, HEATHER, LUKE AND Robert returned to the laboratory feeling well refreshed and in good form. And Luke remembered the last proposition, to the effect that the meat might have to be cut for analysis in order to determine why bacteria had "oozing substances" coming out of them, and why they could not eat the "delicious" morsel of steak.

"Well, let us proceed now to determine bacterial status from analysis of the unspoiled piece of steak," said Matt. "This time let us work together so as to save time and have a more detailed report from the experimental process. We can take turns at both examination and note-taking. How much of a piece of the meat do you think should be cut, so as not to disturb or disrupt 'initial conditions,' for, we would not want the meat to begin to spoil, rot or degrade, right?"

"Wow, Matt," said Heather, "We had not thought about that. But it is a possibility. In any event, keep your minds sharp and observe, write and take note of all data, even the minutest of details, for they may have great import for future research—especially food safety, conservation and storage."

Robert and Luke agreed on a small piece of meat for cutting, with the approval of Matt and Heather. And the experimental process continued in order to determine the cause-and-effect relationships that have undergirded the "suspended unspoiled state" of cooked meat for almost 96 hours.

As soon as the meat was cut, they heard very faint, almost inaudible "crackling sounds" as if the meat was undergoing some unknown chemical change.

"Are you hearing this?" asked Matt. "Please make a note of it."

After two and one half hours of investigation, evaluation, examination and analysis of the meat morsel, Matt gave the signal that it was time to stop and gather together for reporting. It was already 5:30pm on that Saturday, and they were ready to finalize the research, having made sure that no detail was left unobserved or unnoted.

"Wow!" exclaimed Robert Tillkamore. "Heather, would you please smell my fingers, I mean, my gloves. No. Seriously! Luke, Matt. You too! Smell my gloves."

They approached Robert, not knowing what to expect and with a sort of muted caution, common to approaching a wasp whose sting one wanted to avoid. After they all proceeded to prudently smell his gloves, they all had a common thought.

"Matt," said Heather, "those odors, they are things we know and have and utilize all the time."

"Yes, I thought so too," said Luke. "The bombardier beetle is an amazing little insect with 'excretion pellets' that combine bio-ingredients to produce such smells. And when the beetle sends a discharge against a predator, it makes a small 'popping sound' as the chemical ingredients come into contact with the atmosphere. Hence, the little 'crackling noises' we heard, Matt, as the steak was being cut. Do you remember what these recombinant ingredients were, in contriving such a smell?"

"Of course," replied Heather and Matt, almost simultaneously.

"Peroxide," exclaimed Matt.

"Ammonia" retorted Heather.

"Acetone, also" enjoined Luke.

"Could these be our 'antibiotics' or so-called 'house-hold cleaning supplies', that paralyzed bacterium motility and caused 'oozing' out of their inwards from the disintegration of their cell walls?" enjoined Luke.

"Friends, this is an awesome discovery. The combination of an electro-magnetic field engendering gravi-metric molecular effects upon the shed materials, causing them to enact the release of chemicals in the steak molecules that were inimical to bacterial growth and survival, while at the same time sealing and preserving the molecular integrity of the meat morsel itself. That is an amazing mystery!" said Matt.

"But how come only your own gloves, Robert; Why not ours also?" asked Luke.

"May be the time factor," replied Robert. "I was the last one working with and touching the portion of meat we cut from the whole. Perhaps, the molecules released these chemicals upon some special stimulation from my instruments; perhaps, it is the ambient temperature and pressure in the lab. Perhaps, in that shed, the meat "got zapped" with something that's now dissipating. Perhaps I just happened to have accentuated an ongoing process that only needed the right kind of triggering in order to climax to its final conclusion."

"How is the meat itself? Do you notice any change since we broke 'surface tension' so to speak?" asked Heather.

"Well, the piece under the microscope smells more like a bombardier beetle's 'little gas grenade," said Luke. "These gases may have been formed in the shed due to inputs from solar radiation such as ultraviolet and possibly other rays, and electro-magnetic catalysts from 'dry lightning,' as the lightning pole radiated its enhanced field of influence right into the shed where various substances, materials, and liquids may have reacted to produce such an event in the total complexity of its experimental details. And since the meat was the only 'organic container' or 'bio-receptacle,' so to speak, for lack of a better concept, it attracted these molecules that then infused it with reactive ingredients within cellular operations for peroxide, acetone, and ammonia retention."

Matt enjoined, "Is it not in phospholipids or organic fatty acids that storage of certain chemical compounds take place in the human body? Well, beef is a very fatty meat. Its peculiar molecular structure and composition might be accountable for the formation and sealed retention of these 'aromatic molecules,' the tissues being a convenient bio-receptacle."

Robert added, "Yes, from your analyses then, there must have been an ensemble of molecules or 'atmospheric aura' inside that shed so "extra-terrestrial," so to speak, as to engineer the most unlikely kind of gravi-metric and chemo-biological effects, which could have affected even perhaps humans and animals."

"It might have been also during that time that a "vacuum" was created within the sealed bowl, due to pressure and heat differentials engendered by the electromagnetic field, hence, permitting the chemical reactions to take place, forming peroxide and ammonia, and even acetone—the phenomena within the sealed bowl itself having contributed to the 'sealed meat molecules' causing them to 'lock in' the newly formed agents within their atomic structures."

"What if a living person were in that shed at the same time? We have no way of knowing what the effects could have been,—perhaps contra-indicative; or perhaps the human body is engineered in such a way as to 'process' these field occurrences in an innocuous manner that would not have disrupted bio-immuno and electro-metabolic processes."

"I just thought of something," said Heather. "The Trinklungs did open the bowl for a preliminary examination. Then, why would there still be a vacuum, after resealing it in order to transport it to you, Luke? In other words, why would a vacuum persist even after the 'initial conditions' were disrupted"?

"Perhaps resealing the bowl also restored the 'initial conditions' to a certain extent. Wouldn't you think that the meat itself was generating this vacuum due to newly emerging transformations in its intrinsic and environmental properties, such as in pressure, temperature and chemical composition, all, being generated by the steak's morphologic inner-molecular sealing dynamics? These processes are mysterious and multi-causal, to say the least," replied Luke. "Remember, we could not smell nor detect any chemical scent before the meat was cut," he continued.

Matt enjoined, "The greater portion or original morsel of meat, I believe, is now showing some signs of decay externally, with discoloration noticeable whereof the piece of meat was cut from it. It is disintegrating and 'drying up.' "Surface tension," so to speak, has been broken—it's an open 'prey' to bacteria now, lacking a better analogy, for disclosing 'inner molecular secrets.' Should I put it closer to my nose and smell it too? I think I will . . . Yes, there is a slight odor not present before, akin to ammonia-peroxide-acetone combination, which might account for keeping away parasites and bacteria that could have engendered spoilage."

"Heather's concept of 'surface tension,' I think may have had something to do with the timing of the release since initially these ingredients might have had greater intensities, densities and focused concentrations, thus, killing bacteria and parasites via enzymatic action that destroyed their cell walls and caused outward excretions from their inner organs, so to speak," said Luke.

Robert caught on, "'Surface tension' explains why the meat morsel, as a whole, kept together as an 'energy unit.' But what would explain the reasons why we could not perceive or sense these chemicals before by smelling them—and how come once we smelled them we could identify them right way, even without microscopically looking at and detecting their molecular structure and composition? The molecular "vacuum" had sealed their fragrances so tightly as to prevent detection. We had to cut the meat, and then, the process unfolded so that the 'aromatic' or 'odor-filled' molecules could then burst and release chemical 'fragrance pellets' for us to sense with our nostrils."

"I posit to you, that, because of solar events and electromagnetic field processes inside the shed, certain atomic inputs affected the meat morsel at the particles level—creating an 'inner strong force' in the nuclei of its constituent atoms, hence, keeping peroxide, acetone and ammonia molecules in the tightest formation and configuration, and, proving why we had to cut the meat first, before, we could smell ammonia, acetone, and peroxide 'aromas.' And that organic 'strong nucleic force' once broken, triggered the fragmentation of molecular chemical bonding for ammonia, acetone and peroxide atoms, and at the same time, engendering the beginning of the very spoilage process that was heretofore postponed."

Matt added, "So this portion of cooked meat, stored since Tuesday evening, we could say, lasted until Saturday evening, without any remarkable degradation or spoilage, without exuding any smell or odor, as a 'self-contained, whole-energy system.' We speak that way due to certain laws of Physics that have been triggered because of solar radio-gravitational activities of late. That's about 4 days or close to 96 hours of Conservation, while 'Terminal Entropy' was efficiently thwarted. However, the meat was no longer edible or 'consumable,' it had no electro-conductivity and appeared 'dead.'"

"Your analyses suggest that, due to solar gravitational transformation dynamics occurring for the past two weeks, there is also occurring trans-morphologic processes in organic matter; and that not only molecular dynamics but also atomic mechanics are engaged, because of electro-magnetic forces and binding energies interacting with bio-organic matter, in this case, the portion of meat stored in a tightly sealed plastic bowl. This is Physics interfacing with biology at the deepest core of physiological chemistry, the rarest occurrence in the universe of controlled experimentation."

Robert continued, "And the morsel of meat did not have any odor or spoiled smell, not until operated upon in the laboratory; after which it was discovered that the meat portion developed a kind of 'protective tension,' accruing from molecular morphologic transformations that led to the 'inner-force-conserved,' self-contained, production of acetone-ammonia-peroxide-like compounds within its cells. For, the meat's constituent elements being mostly protein cells made up of amino-acids joined by peptide chains that contain many elemental substances such as Carbon, Hydrogen, Nitrogen, Oxygen, Sulfur, Phosphorous, Iron etc. . . , thus precipitated the formation of the chemicals via sealed molecular reactions,—such as ammonia, having the formula NH_3; acetone, with the formula C_3H_6O; and hydrogen peroxide, of course, a compound constituted of Hydrogen and Oxygen whose formula is H_2O_2, being a chemical possessing antiseptic and propellant properties,—which climaxed into their being finally stored within the steak's fatty acid tissues."

"For example, acetone itself is commonly used as a solvent, which might explain the breakdown of bacterial cell walls thus engendering these 'oozing' substances from their inward parts. All these recombinant chemical properties, along with 'surface tension,' and sealed molecular dynamics, worked to insulate the steak from bacterially-generated enzymes involved in organic degradation."

Luke enjoined, "Compounds, which, to our deductive surmise, once consumed or released, were responsible for warding off parasitic and bacterial predators that could have accelerated spoilage and degradation of the beef steak to a state of 'rotten meat.' And the ammonia-peroxide-acetone compounds, analogous to bombardier beetle 'gas pellets' are constituted of the same ingredients utilized in household cleaning agents for disinfecting and sanitizing people and spaces; however, the 'surface tension,' once broken or released, no longer ensuring protection from spoilage, decay or degradation, accelerated the steak's disintegration."

"So, where do we go from here, friends?" asked Luke.

"Let's talk to the Trinklungs after which, I believe, we should hold a press conference in order to disseminate the contents of our findings, details of which might be useful to farming families, food processing plants, agri-business concerns, chemical manufacturers, and other affected groups and parties with an interest in safety and sanitation, in general, and in food conservation and storage, in particular," said Matt.

"However, Matt," added Heather, "we should consider beseeching our fellow citizens and neighbors not to attempt to reproduce these electro-magnetic and molecular events from their basement or garage, explaining that things just happened to have developed in this manner due to unusual and irregular solar activities that affected Earth weather systems, magnetic field and surface gravity, etc . . ."

Luke agreed, "Yes, that was a special event that took place under specific experimental conditions, which originally, were not intended by human will, but which occurred as if 'randomly,' or 'by chance,'—that is, in the absence of humanly prepared lab conditions, the full and complete parameters of which, we do not know nor could we know. And we could also entreat them to understand that research results ought to be utilized in a manner that alerts every one to pay attention to surroundings and environment in order to detect uncommon processes or events and report them to the proper people, as appropriate. For example, this is good class experience for the high school where I teach, especially for Marc, the student who came across this discovery."

"Well, Matt, any final words?" asked Robert.

"Yes, please, all of you, make your personal reports from notes and observations, with any additional inferences and deductions, and mail them to Luke at the high school. Luke will then confer with me, as to final version, so that at the press conference we can all have the same rendition for reference purposes, in addition to our own particular expertise at confronting this type of research challenge. I will analyze all your individual reports to summarize them into a coherent, well-integrated thesis, taking care to fit all details within the paradigm we have all uncovered from gravitational solar dynamics in addition to Earth gravity-sensitive responses, as manifested in specific local processes, such as the 'meat situation' at the Trinklungs. Good job, every one."

Luke could not contain his enthusiasm, "Thank you all of you for helping out an old friend. I am forever grateful. I could not have conducted this lab examination all by myself. Your inputs were vital to our reaching conclusive results which will be useful to a lot of people as we all have to deal with the new changes brought about by transformations in solar system dynamics."

After exchanging greetings, they parted, promising to keep in touch in light of any development having bearing on their common project.

*   *   *

11

O N THAT DAY, IT appeared to be just another ordinary morning. Yesterday evening was hazy, cloudy, and humid, diffusing a "still-at-rest" tone in the air reminiscent of pre-tornado striking conditions. Not even a flutter in the leaves of nearby trees, not even a quiver in the bushes bordering the Trinklung's property! But, nothing significant took place, weather-wise.

That Monday morning, however, was almost ominous of a coming catastrophe. People went about their business in the usual way, keeping a positive attitude, yet not knowing what the solar system will do next, and relying upon news broadcasts and other informative processes to remain current on solar gravi-dynamics.

Every thing was quiet around the city high school. It's a building composed of three adjoining structures, the central one, being the main entrance to the school. It was just about time for students to start arriving.

A group of students leaving the bus got ready to enter the high school building when one of them, Suzan, exclaimed, "What's going on with this flock of geese? Look, they usually fly high in the skies while following a fixed flying pattern and seldom make any stops in cities. But why would they fly so low?"

The students looked and saw the whole flock of geese flying into the middle structure all at once, as if they had lost the capacity to navigate through an environment constituted of solids like trees and mountains and buildings.

They all actually crashed into the central structure, analogous to the way in which a whole flying team crashes due to an error committed by the lead flyer during an air show. Their guidance system, or the way in which they sense their environment in migrating from one region of the globe to another, must have gone askew. But what could have caused these birds to do that? Sometimes, whales "beach themselves" also, with no apparent cause or motive. However, these school structures were not that high and did not even possess flight signals for aircraft alert, as high communications antennae or industrial smokestacks do.

The principal of the high school was alerted to the event and the maintenance staff was instructed to remove the birds from the property, but not before having reported it to the county agency serving as a clearing house for "unusual incidents." The county agency contacted university scientists to report the event. They, in cooperation with civilian and federal space monitoring organizations, then proceeded to review solar activity and Earth eco-system reports in order to determine the possibility of the next gravitational process that could affect the Earth magnetic field actions, core electro-dynamics, atmosphere, and gravity-characteristic functions—for, something had affected those birds's ability to navigate, in a way or form that is yet unknown.

For several decades now, since applications of the Theory of Relativity had been extended and expanded to the universe as a whole, scientists have engaged in diverse elaborations regarding solar system "death" or solar "Terminal Entropy,"—from so-called "yellow dwarf star" to "red giant," to "white dwarf" etc . . . ,—but with such professional nonchalance and academic levity as to not have aroused any serious concerns in people simply engaged in "making a living" or providing the wherewithal of daily life to their families. However, events in the last few weeks have changed this state of affairs somehow as the population cares about, more than ever before, any news concerning changes in Earth and sun parameters, such as star spectral emissions, Earth atmospheric pressure, geo-temperatures, solar winds and flares, tornadoes and hurricanes forecasting, typhoons and tsunamis predictions, coronal mass ejections, plate tectonics, earthquake detection etc . . .

"Science-news," so to speak, has gained top priority with people every where on Earth. And more than ever before, the onus has been put on scientists to "deliver" to the best of their learned capabilities, the most accurate information in light of the knowledge and experience they have acquired all along their professional and service careers.

*   *   *

12

A T 1:00AM, TUESDAY MORNING, the President of the United States and specific members of Congress, were awakened by NASA employees, with most threatening news coming from observations of the solar system by various solar observatories. The news media in all its variegated forms were alerted not to release any data as of yet in order to avert mass panic and inordinate fears as to upcoming consequences, to which they adhered generally, in not speculating on the specific description of the unidentified object that was racing towards the Earth. It did not take long for appropriate federal agencies to inform state and county counterparts in the nation whose responsibilities involved emergency preparedness, readiness and response.

All concerned parties were summarily informed that the object approaching Earth orbit will enter its ecliptic plane in approximately 19 hours, given its apparent mass and traveling velocity, to pass near Earth orbit, however, approximately 3 hours thereafter.

Questions arose regarding the space-body's size, shape, surface texture, spectrographic and thermal signatures, which could not be directly evaluated at that time.

The President and others learned that a preliminary assessment ventured the hypothesis that the approaching object might be "a comet" due to the structure apparently trailing behind in its wake—in the likeness of an "ice tail." Many inquired if it could be "Halley's comet," with an answer specifying that "comet Halley" was not due until 2061—because it had a galactic orbital cycle of 75 to 76 years, its last appearance having been in 1986.

"Then, what else could it be?" asked the Speaker of the House. "Thus, according to the information we've just received, this 'thing,' whatever it is, will be near or approaching Earth orbit at about 11pm today, this evening, tonight."

"Could it be a big asteroid?" inquired the President.

"It is hard to tell at this time, Mr. President," answered NASA personnel. "However, as its approach becomes less and less distant from the Earth, we might have a better determination."

"Can we afford such a luxury given that it could hit the Earth and especially, the United States?" thought the Director of Homeland Security Department out loud.

"Scientists are working on it as fast as they can," added NASA personnel, "even the European Space Agency and other space observing concerns such as in Japan, Canada, Russia, France and Great Britain. It's now 3:30am, Tuesday morning—the greater numbers of people on the Earth are peacefully ensconced in their beds, sleeping, and hopefully, they will have had made a determination before day break."

Evidently, such "big news" could not be "kept under wraps" for too long—the only thing remaining to be discovered was the identity of the object. A phone call to NASA's newly confirmed Office of Space Exceptional Circumstances, from SOHO—Solar and Heliospheric Observatory—redirected towards the White House, informed all present that "scientists had tentatively agreed an object or space body 'in the likeness of a comet' had been approaching the Earth for the past 6 hours; and given the limits of our observatory instrumentations, near-Earth orbit technologies could not have detected it but a couple of hours ago. The estimated 6 hours was due to our knowledge of the orbital spans of the solar system and the Milky Way Galaxy."

It was now 4:00am, the President and Congress, in consultation with various scientific establishments, state and county officials, decided to keep students at home on that day. The media was contacted—with caveats against "wild dramatic speculations"—in order to make announcements that school was going to be closed due to unscheduled and unforeseen events happening at various school districts in the country, requiring all districts participating therein.

They also thought it best to schedule a public event—which they would have an estimated 6 hours to prepare for—that underscored and emphasized focused attention upon the analytical and thinking capacity of human beings in the presence of potential or impending disasters for purposes of preparedness, reasoned judgment and functional readiness, in the likeness of a symposium of qualified scientists and concerned parties, that would also be open to the public at large, to take place within the greatest indoor sports stadium they could find in one of the Midwest States, right at the center of the country.

Kansas appeared to be a good candidate for such an event. After all arrangements were made through appropriate channels and due process of law observances, the symposium was scheduled for 10:00am that Tuesday morning. Congress will have had convened at 7:00am in order to enact pertinent legislations and monetary appropriations that would support attached expenditures or subsidies. All participating parties will have had been informed, including leading scientists with respective expertise in fields of application, such as astrophysics, solar meteorology, matrix cosmography, Earth-sciences, solar plasma dynamics, bio-physiology, galactic mechanics, and electro-microbiology etc . . .

Every one was aware of the time constriction requiring the most extensive application of emergency procedures in order to prepare for an event, the consequences of which, too dire or dreadful to even consider. A space object or body of such gigantic size hitting the Earth or even "grazing" its atmosphere would mean a cataclysm of such proportions as to overwhelm the sensibilities of most people present—they dared not imagine such a horrifying destructive holocaust.

They then agreed that the widest possible form of public participation and openness to the media were necessary in order to avert panic and hence, extreme "survivalist" mentality that would trigger hoarding of food, building materials, and fuel supplies, which would clog circulation in small towns and cities around the nation. It could not be "business as usual" but at the same time, precautions had to be taken to prevent conditions from emerging that would precipitate even greater unwarranted disasters of human doing. For, no "catastrophe" had taken place yet—at least, it was their hope that they would be able to prepare for any eventuality that would threaten human lives and the stability of civilized living on the Earth.

*   *   *

13

THAT WAS THE GREATEST gathering ever assembled in the United States of America—even the playing field of the stadium was full of people who brought their own lawn chairs in order not to miss participating in such a momentous event put together in so short a time due to the emergency situation evoked by the space object.

Organizers of the event had been well briefed on needed technologies and public announcement systems; and their logistics experts, highly motivated by the circumstances of the moment, had taken great care to make provisions for all accouterments and equipments; not even the portable water closets were forgotten.

The "master of ceremonies" wasted no time in welcoming the whole audience, with some information regarding the purposes of the event as convened from Washington DC in cooperation with state, county and local authorities, and civilian organizations responsible for structuring the conference. He then introduced Peter Barlotuk, the highly acclaimed astrophysicist from Illinois, to begin the symposium. Various microphones were strategically placed within the vastness of auditorium space for participants in the audience who had questions and comments to share in the course of respective presentations by experts and scientists scheduled to speak.

"As you well know," began the renown scientist, "for a few weeks now, there have been various manifestations of unusual or uncommon solar gravitational and radiant activity affecting the Earth in multi-various ways that caused oscillating variability in its magnetic field, gravity quotient, geo-atmospheric motions, weather systems, core pressure differentials, tectonic activity, photosynthetic processes, organic and molecular dynamics, immuno-hormonal metabolic effects, etc . . ."

"It appears that the Earth is not the only thing being affected by such new solar transformations—now there is something, an object or body approaching the Earth and it is believed to be a comet-like space body, due to an appending structure believed to be an "ice tail," common to 'comet Halley.' Because of this specific event, preliminary estimations indicate that cosmic disturbances are occurring in space regions beyond our solar system, precipitating phenomena that could have affected the motion patterns of stellar and cosmic bodies that heretofore had specific orbits within the galaxy, but not within the solar system itself."

"In other words, these occurrences may have their nascent beginnings or interstellar genesis in the outermost periphery of the Milky Way Galaxy wherein the sun and the whole solar system is traveling or moving. These extrapolations have been made in case that this approaching space body is 'comet Halley' which is due to return in 2061. If it is not, then it is a space body or object akin to a comet which has never been recorded or seen before in human history. In any case, it is an object, the orbital path or trajectory of which, has been disturbed—it was or may have been 'thrown off course,' so to speak."

At that time, Peter Barlotuk paused to introduce the next speaker, Philip Karbidek who continued, "It has been agreed that these perturbations might have begun beyond the heliosphere, deep into the interstellar medium, because the object presently approaching the Earth has not been predicted or forecast by any known physical law of the universe. It could be a 'glitch,' so to speak, in cosmic operations, due to Relativity ramifications of extremely warped space occurring because of unknown gravitational activities at the deepest outskirts of the Milky Way Galaxy—beyond the ability of any our technological instruments to detect."

"The sun rotates upon its own axis in every 27 days and also revolves around the galaxy in conjunction with the Milky Way Galaxy rotating around the Andromeda Galaxy, and so forth."

"In addition, the sun undergoes a magnetic field polarity displacement every 11 years, which is recast in another 11 years as a complete magnetic field polarity change—therefore, constituting two cycles, every 11 years and every 22 years. Magnetic field polarity change, taking place every 22 years, is usually accompanied by great turbulence that exacerbates every radiation-sensitive process in the solar system."

"The whole universe is in continuum motion—things revolve around each other: planets around stars, stars around galaxies and galaxies around each other. Motion is generic to continuum Space-Time in which we, in this solar system, also find ourselves."

"The object approaching Earth is massive and is traveling at a velocity that will allow it to approach the area covered by Earth orbit around the sun, at approximately 8:00pm, to then penetrate within proximity of Earth orbit itself, at about 11:00pm. There has been no known displacement of other planets or asteroid dislocations from the asteroid belt, and no motions were detected from space bodies or objects within the Kuiper belt. However, we already know the path of 'comet Halley,'—the trajectory being followed by this object is not that of 'comet Halley,' though in material appearance, it resembles 'comet Halley' somewhat, due to the "ice tail structure" seeming to accompany its body."

"The object and 'comet Halley' appear to be comparably of the same mass—or the object might be of slightly greater mass due to heliospheric acceleration in its perihelion travel in relation to the sun—per gross estimations made by matrix cosmographers trained in astronomical measurements. That means, as the object is racing through cosmic space, it has been accumulating energy which, upon impact, will reverberate as the greatest exponential shock wave event."

"Thus, the crucial concerns at this time are whether or not it might hit the Earth in direct impact; or if it will simply pass-by Earth at a safe distance; or if the object will even approach Earth to 'graze' the atmosphere, hence, making contact with the magnetosphere and the ionosphere."

"If it hits or 'grazes' the Earth, depending upon its 'angle of attack,' and the accelerated angular momentum as preserved by conservation factors, the impact could shake the Earth with the greatest jolt that may even cause it to "fly out of orbit"—temporarily that is, to then, be "jolted back" into its regular trajectory around the sun, assuming no decrease in heliospheric binding energies that constitute gravi-metric motion variables for revolution and rotation."

"The solar system works in operation like a stretched rubber band, or compressed spring, and if it's cut at any part, or bent in any area, that will cause reverse contracting-dilating reactions in the space covered by the gravitational fields of influence of each planet, and especially of the sun, whose great mass is already the most potent of all pressure-forces within the heliosphere. System equilibrium will have been disrupted and that will 'shake things up' until dynamic equilibrium is again restored. It will be like the Earth undergoing the most extreme 'heart attack;' it will pulsate, vibrate, and contort in the most convoluted motion-vectoring ways, the multi-various effects of which, we cannot even begin to deduce."

"Equilibrium is never static or inertial, except in the case of a simple balance or scale both sides of which contain the same amount of matter or mass. However, Earth events and universal processes are generally dynamic, multi-factorial, and multi-causal—embodying motion, change, productive activity, property differentials, etc . . . After all, the solar system is constituted of a star, eight planets, and Neptune being contoured by more than a dozen natural satellites, each presenting its own gravi-metric signatures."

"The solar system is a rather complex entity, but also a place, where position, motion, velocity, and location-specific gravitationally-sensitive differentials abound. For example, universal processes represented by computational equivalencies, such as in $f = ma$, or $E = MC^2$, where one side of the equation might hold more than one variable, their relationships are configured in many complex ways that include proportional differentials. Such differentials are commonly indicative of operational determinants confined to the initial properties and emergent characteristics of specific processes occurring. A variable on one side of the equation might have to be equated with two or more variables on the other side of the equation in order to achieve equivalency, while relying on constants or conversion factors that facilitate congruency, coherence and integration."

"Consequently, Earth dynamic equilibrium depends upon not only 'initial conditions' of Creation, but also upon gravitational adjustments or interchanges developed during the lifeline or historical timeline of the planet within its cosmographic environment that are replete with conservation activity and entropy-driven reactions conducive to its 'negotiating' cosmic, galactic and solar inputs."

Hernbank Vultig, earth sciences specialist added, "We cannot begin to contemplate consequences of a direct hit by such a massive space object. Many present here, perhaps, have already experienced the shocks of earthquakes and have an idea of gravity-effects upon things and people."

"Things would not only fly out of place to collide with bigger things, but friction between metallic components could ignite fires; dams could break to overflow rivers to cause intense flooding; the landmass might be jolted to such an extent as to move the oceanic hydrosphere into the creation of great tsunamis that would engulf and inundate coastal areas with great diluvial volumes of water; tornadoes could develop due to geo-electro-atmospheric transformations of molecular gases; volcanic eruptions could surge through Earth surface to incinerate vast regions of the country; earthquakes could shake the Earth from increased plate tectonic displacement; oceans could boil to evaporate such great volumes of condensed water as to then flood the earth with long periods of rainfall; the polar icecaps might melt to drown oceanic coastal regions under sea water; the water table could be disrupted to dislodge the emplacement of skyscraper foundations and thus uproot them from the Earth causing them to fall from their vertiginous heights; trains traveling at the time with hazardous chemicals could derail and explode into toxic fumes that could cause millions to die, etc . . ."

"Tragic effects will vary depending upon how far the whole Earth is jettisoned deeper into 'curved solar space' or 'bent planetary spheres of influence'—either at aphelion or at perihelion—hence, its destiny resting upon kinetic determinants of placement within solar system cosmography. It could either collide with some planet or planets, or be thrown into a path such as that it will twist and turn aimlessly until gravi-radio parameters stabilize."

"The shock wave will be great; however, since the space object mass and size are still smaller than Earth mass and size, then it is hoped that destructive effects will be restrained or controlled by gravi-metric determinants which have protected the Earth and other planets for thousands of years from orbital decay and magnetic field degradation."

"We could not know in advance nor project in anticipation all the ways in which Earth life-support systems might be affected but all these processes and events are possible outcomes when or if Earth dynamic equilibrium is disrupted at such a deep level of cosmic crash."

As Hernbank Vultig was motioning for Paul Mirdewvell to speak next, he noticed the sky has been darkening and clouds have been accumulating in the area, with the sun still shining above cloud cover. He did not think it would rain. Suddenly a loud noise, the biggest lightning strike took place in the vicinity, about 4 miles from where they gathered. The flash was so luminous and intense that it illuminated the whole region, even near the stadium whose roof was partially made of glass, and through which they all could see the skies shine with "strobe lighting." Two more strikes, and then, a huge explosion, yonder in the distance; and then things calmed down so that they could continue. It was considered "dry lightning" because no rain was forecast for the day; however, they could sense that something else happened.

It was summarily reported in the mass media that a gas station within the 4 miles radius of the lightning strikes was hit, thus igniting the subterranean gasoline storage tanks, which exploded with great force. People there carrying on business were hurt; and many were being taken to nearby hospitals by ambulance. Fortunately the underground explosion was restrained by concrete containment casings, which allowed many to escape from being burnt to death. Gas pumps that were combusting shot upwards, hence, diverting the flames towards the skies.

At the stadium, there was the "silence of the graveyard" for a few minutes—for lack of a better analogy. Since everyone was on guard for anything to happen there also, after collecting himself, Paul Mirdewvell proceeded to begin his *exposé* regarding the latest developments.

"We are hoping that the sun will not behave in any unusual way such as spewing forth solar flares, solar winds, or coronal mass ejections that are of greater intensity or pressure differentials than usually emitted by its plasma convection activities, in order that damages caused by the space body, if any, will not be exacerbated as to their destructive range of impact. We are asking all of you to keep calm, a 'cool head,' not to panic, but to rather prepare and be ready, in case such a catastrophe occurs."

They were ready now to take questions or listen to comments from the attendees who made the special trip for this meeting. A man, in his forties, Joseph Certill, stood up near a microphone and asked, "Could not the U.S. Air Force or the U.S. Navy target this object to disintegrate it into little pieces with a missile aimed at it from the Earth, at a distance safe enough so that debris would not fall into Earth atmosphere or hit the Earth?"

Dennis Krybonzak, was asked by Peter Barlotuk, to answer this question because Krybonzak was an aerospace engineer and medical doctor. He proceeded by saying, "It is a possibility which has not been taken into account and yet not entirely impracticable. However, the space object has not yet been definitively identified in terms of geo-composition, chemical structure and radioactive matter, if any,—whether it carries materials lethal to human life and Earth life-support systems has not yet been assessed; and therefore, due to these unknown variables and other yet-to-happen factors, such as potential solar activity and disruptions from other planets, that option will be kept at minimum consideration, until the trajectory of the space body can be more surely ascertained."

"In addition, it has been assumed from pictorial analysis that its path is not uniform. Its motion is erratic and its pattern stochastic, or "random" in direction, due to oscillating gravi-metric forces and motion-pressures impinging upon its center of mass—which makes it very difficult to mathematically formulate or quantify its trajectory and other crucial parameters."

"In short, 'we could miss,' which could aggravate the already tragic situation we are now facing. The space-body's angle of approach is not perpendicular to Earth rotational axis, thus forming an acute angle with the Equator, due to the Earth 23 degree tilt—that is, if we were to track it from a U.S. location as in the State of Florida."

The meeting lasted 3 hours, ending at approximately 1:00pm, which permitted all attendees to go eat lunch as they pondered explanations and logical inferences from information received. The options were to either let things take their course or strike the space object with a missile at a distance far enough from the Earth so that debris would not fall back on the Earth. But so many other factors and variables were co-determinants in the equation that no decision could be made at that time.

*   *   *

14

UNBEKNOWNST TO THE PEOPLES of the Earth who could not inspect the internal dynamics of the sun's core, the thermal radio-active substrate of the solar inferno has been contracting as plasma particles compressed into complex micro-molecular magnetic tornadoes thereby increasing solar temperatures to such great intensities and plasma molecular particulates to such great densities, that convection mechanics have been suffering delayed chain reactions— hence, the sun temporarily "blacking out," so to speak. But when the core pressures became intolerable as to exceed the threshold of plasma binding energies and condensate densities, convection was again triggered with the most violent of momentum propulsion forces rushing outwards from core center of condensate explosions fighting the counter-force of gravitational contraction.

Great thermal energies were compressed at their proton-nuclei levels while neutrinos, positrons and cosmic rays escaped to the convection region to release tremendous amounts of energy as they were propelled from one energy level to another towards the corona, to only be re-ejected inwardly towards the sun's center of mass. Due to these extreme mass density differentials, one could only imagine the immense pressures and temperatures being churned during these centripetal compressions.

The sun is comparable to a huge fusion reaction mass experiencing continuous pocket fission ejections that are then violently re-compressed towards its core, while only the lighter particulate emissions manage to escape from its immense gravitational electro-radiating magnetic field. It is analogous to an immensely compressed-explosion exhibiting both centripetal and centrifugal forces, pulling towards and pushing away from its core-dominated center of electromagnetic mass, in constant counter-vectoring motions, all at the same time, with neither force ever achieving dominance over the other—hence, its usual operational dynamic equilibrium as a bright, shining, radiating star.

Examinations of the sun were diverted towards observation of the space-object approaching the Earth, a catastrophe of galactic proportions rather than a problem originating from solar system operations. Media announcements at 7:00pm on Tuesday, only confirmed that the object was of greater mass than originally thought, and hence, eliminating all speculations that it could be "comet Halley" resurrected. Additionally, space observation centers around the world reported to have seen electrical activity in the form of arcs on the space-body or object as if it had capacities for absorbing and exuding electromagnetic energy, including an electric field and a magnetic field, which, in the worst of scenarios, could interact with Earth magnetic field and electrified atmosphere, to engender the most destructive forms of turbulence. Were a missile to have been deployed, there could have been electro-connectivity between the Earth and the object which could have caused atmospheric or magnetic field disruptions resulting in the missile exploding prematurely in Earth atmosphere.

People were advised to anticipate impact in terms of effects akin to tornadoes and earthquakes, and that there was a relationship between all shockwave patterns, whether originating on Earth or from outer space as in a potential collision with a cosmic body, this analogy implying that preparations could be made in anticipation of possible damage and destruction ensuing thereof.

Suddenly, the sun convulsed with great unrestrained emanations engulfing the solar surface with streams of plasma fire. Pluri-potent or omni-directional packets of particulate magnetic fields that both attracted and repelled at the same time,—pocket fissions and fusions—rushed from the sun's center of mass to the dipolar magnetic field, short-circuiting its conductivity to then re-penetrate the inner solar atmosphere, arousing tremors from within, with the creation of dielectrics—regions or spots that, though they do not conduct electricity, however can sustain an electric field. These dielectrics were suctioned back, as in a great vacuum and re-absorbed to again re-actively radiate binding energies from the core's extreme plasma fuel temperatures and pressures; as they were forced to fuse, while countless contractions and repulsions traversed the radiating area, they re-emerged within the convection region as spectral emissions, catapulted towards the coronal circumference.

The sun's hydrostatic equilibrium had been disrupted—its convulsive contractions and expansions shocked it while attempting to sustain wholeness within its huge magnetic field; its magnetic field could have collapsed, except that its mass is so weighty as to compel electro-magnetic polarity compression. This compressive-implosive force had an exponential effect upon inner activity for plasma fusion chain reactions, from Hydrogen to Helium, while the dilatory-explosive forces engaged the nuclei of heavier elements like Iron, Carbon, Oxygen, Silicon, Nitrogen, and Sulfur. The convection region served as the 'recycling template' within the dimensions of which particulate emissions released their tremendous energies from one energy level or orbit to another, while being subjected to two simultaneously applied counter-forces—impelled towards the core and at the same time repelled towards the corona for cosmic escape.

Such fusion-fission counter-interchanges as contained by an immense magnetic field of force, allowed a certain amount of cooling as the sun released cosmic energy throughout the heliosphere. However, it appeared that particles jumping from one energy level to another re-synthesized atomic energy unto transmutations of thermal radiation while avoiding the convection region,—during which they re-absorbed the radiation that was intended for release, light and heat energy, hence, contributing to the sun's instability as a whole-energy system direly needing to restore dynamic equilibrium.

At 10:00pm, it was announced on television and radio, that the space body, half the size of the Earth Moon, could be seen from the Earth. The Moon is a quarter of Earth size and the space body was half the size of the Moon—so the space-body was about one eighth of Earth size. That was still a mass big enough to be reckoned with, but not so extreme as to cause irreversible Earth perturbations and disruptions, such as permanent orbital dislocation or decay.

At 10:20pm, the sun's "safety valve opened," and triggered successive explosive coronal mass ejections so big as to evoke the fluidity of "plasma rain." The convection region was attempting to claim its share of plasma processing. These eruptive outflows accompanied by electro-conductive radiant flux, covered the space object with plasma condensate and radioactive particulates, extending in their surge to reach the Earth at the same time.

The space body spun around violently when hit by plasma ejections as the "plasma rain" deluge connected with the Earth dipolar field. The Earth responded with radio-sensitive atmospheric strata reshuffling their gaseous structure and composition, causing lightning and hail formation to rage on.

The object appeared as a huge red ball, rushing towards the Earth as if gaining in accelerated momentum. When the "plasma rain" reached the Earth, Aurora Borealis at both poles gradually penetrated the atmosphere with a great rush towards the Equator—where they joined at the Equator for a few seconds, there occurred sharp and eccentric electric shocks, unlike lightning, but more like two "live-wires" intersecting each other, while slowly diminishing and retreating, pulsating back and forth towards and away from the polar regions. Some trees, even green ones were scorched to a crisp while many transformers blew and many houses experienced temporary interruptions in electrical current.

At the exact moment that Aurora Borealis "merged," as if choreographed by cosmic forces, the space object, with added momentum, rushed towards the Earth as if to enter into its atmosphere or hit it directly—but bounced back with such a great recoil as to engender a counter-recoil reaction unto the Earth, shaking the landmass with twisting tremors that evidenced in the hydrosphere as violent oceanic undulations. The space object did not have the correct "angle of attack" and thus bounced back into space towards the Moon whose tidal rebound sent the space-body spinning into the direction of Earth North pole.

That engendered a shock or quake of great magnitude on the Moon which "jarred" it from its orbital path for a few seconds, initiating upon its return to regular orbit, tidal forces of such great intensities as to transmogrify the oceans into towering wave structures with crashing powers

against shores, coastlines, beach houses, and bridges etc . . .
The ensuing surge of ocean waters could not be retained,
contained or arrested by shore barriers erected to protect
nearby buildings and structures, regardless of their height
and mass.

And as the space-body bounced back from the Artic, its
electric field made a connecting arc with the Earth magnetic
field, the continuum current of which increased atmospheric
electro-conductivity all the way down to the hot iron molten magma core which reacted with violent overflows—lava
pillows on the sea floor "bulged out" as they spewed forth
from the inner bowels of the Earth to get cooled and solidified by deep ocean waters as they rolled down ridges, valleys
and hills beneath the waves. When the cold sea water made
contact with the hot lava, minute molecules of salted waters
boiled on the contoured surfaces of lava pillows as temperatures rose beyond normal ranges.

Prolific lightning activity above the oceans lit the hydrosphere with such tremendous intensity as to cover the
circle of the line of horizon with a colorful brightness that
erupted to rise into the atmosphere, hovering just below
cloud cover. A fire erupted in the vicinity of the Atlantic
Ridge due to an oil deposit ignited by electrical forces. The
waters above the Ridge now flooded with petroleum that
gushed out of its sanctuary from deep within the Earth
crust, exploded as lightning strikes inflamed its combustive
compounds. That fire will have burnt until its sources of petroleum oil are exhausted.

The jet stream "clashed" with El Nino as they were
forced into a contractive vortex encompassing both hemispheres, creating whirlwinds that were accompanied by
great flashes of lightning, "peals of thunder," hail the size

of golf balls, and rain the droplets of which were so hot as to burn unprotected skin, these two complex forces "losing steam" as they traveled from the equatorial towards the temperate zones.

In Hawaii, volcanoes erupted with a raging violence never seen before, with hot red lava flowing from their "belching mouths," so to speak. People were fleeing from coastal areas to join inland security and safety enclaves prepared in advance for such a catastrophe.

In Southern Italy, Mount Vesuvius, which had been dormant for so long, since it buried Pompeii in AD 79 hundreds of years ago, roared once more to scatter hot gray-dark ashes for a spread of 3 miles, as inhabitants of the region, tourists and spectators ran for cover.

Earth landmass shook greatly as if torn from its foundations, as the core received its greatest "lightning rod jolt" so to speak, from the space body whose electrical megawattage overwhelmed the weak Earth magnetic field. Its great "anchor of gravity" was assaulted even to the "links" of its supreme mass-sensitive "chain," as if its spherical but challenged symmetry could no longer sustain the processes that enlivened its superlative exceptionality as the only life-planet in the whole universe. Space buoyancy was now its greatest ally if only it could "weather" the shock from the space object in conjunction with the sun's own plasma condensate, and cosmic particulate "storm systems."

The atmosphere "smelt burnt" as gases convulsed in trying to accommodate such great amounts of radio-electricity without altering their breathable composition and hydro-forming proportions. The Earth revolted in jarring and oscillating motions and trembled in horrific tremors that seemed to blur the lines between landmass, hydrosphere and atmosphere. People saw sea waves rise from the shores as if they were on a ship sailing in open ocean waters, to experience equilibrium disrupting sensations that caused them to fall on the Earth as in induced "motion sickness." But water attracted water. The waters seemed to be reaching for "gulps of the atmosphere" only to be stopped by their own mass and land beneath whose various higher barricades caused them to crash onto the ocean's surfaces once more.

After such long periods of drought all over the Earth, suddenly it began to rain aversely, so profusely as to make rainwater molecules look like small "cannon balls" due to the great atmospheric pressures and gravity momentum that caused them to crash with such great force upon the ground. The waters appeared to hover for a few seconds above the scorched hot soil to then be suctioned within the Earth with such rapidity as to mimic electrons "flying around" and "assaulting" the atomic nucleus to then be repelled by the proton/neutron strong force. It was as if the scorched Earth reluctantly welcomed this long-awaited miracle because it was accompanied by great upheavals that caused the oceanic hydrosphere to convulse with inundations of the shores where sweet rain water mixed with salted sea water; it was as if crops nearby "cringed" while soaking the oncoming onslaught interfering with their planting and growing, and maturing ripeness, as they "looked forward" to harvesting season.

The shocks prodding the Earth sent it "flying." It is as if the Earth began to wildly race into space—rudderless and chaotic, it seemed to have become "unhinged" from its "gravity buoy" that kept it in orbit around the sun. An anomaly occurred—the Earth was thrown into a course that changed its angular distance around its orbit from its perihelion—it got even relatively closer to the sun. For a few seconds, temperatures rose on the Earth, especially at the Equator. That was a good thing for it to have begun raining—as if the Earth had "anticipated" an excess in temperature increase. "Heat stroke" claimed the lives of a few people whose medical condition was precariously being sustained by hospital personnel who did all in their capacity to save them, but to no avail.

Planets around the Earth began to "react" in disruptive "space bending" and "space curving" resonances and reverberations, their respective perturbations testifying to the gravity of the moment when the sun's tempest threw all gravi-metric variables into recalibration status. What were the other planets to do?

Since the solar system works like a stretched rubber band or compressed spring, tidal effects from the other planets "pushed" the Earth back into place. For, the sun had not "died out" yet; it was still the solar system star keeping the whole entity together with great gravitational might. The sun appeared to be even more powerful in keeping and "possessing" its planets into place around it. Its mass had absorbed so much energy from oncoming interstellar plasma winds that the core responded with great agitation, contortion and convulsion in attempts to "negotiate" these gains in nuclear and atomic fuel that disrupted solar dynamic equilibrium, causing it to "oscillate" between a spheroid and an ellipsoid in form.

The Earth recoiled, rebounded and wobbled back into its pre-designed revolutionary position around the sun; its rotation "got chocked" so to speak, like an old carburetor gasping for air, as it tried to then re-position itself back into its tilted axis of rotation—Uranus orbits the sun on its side; the Earth had to regain its 23 degree tilt of rotational axis for life to continue to prosper.

Finally, the Earth, oscillating back and forth like a pendulum, and swinging back and forth from momentum forces that caused effects greater than Gravity-1, re-clocked itself to reset its motion back into the rhythm and rate of 24-hour rotation and 365$^{1/4}$ day revolution. Its velocity stabilized, the distance it traveled in space as it moved was restored, its motions becoming more uniform, respectively, and its gravity reaching "near-normal" operational magnitude and direction—pliable but wrenching attempts at re-establishing standard regularity!

On the Earth surface, however, every thing, motion-wise and vector-wise, was "upside down." Its poles "got flattened" a little bit more due to increased solar gravitational plasma binding energy releases. The air was raging with uncontrollable storm systems that caused diluvial rain falls and electrically driven atmospheric conflagrations. Many tornadoes were formed and touched land while creating havoc in their path. A sink hole was created in Florida but this time it was gushing out with petroleum oil that got ignited by much-dreaded lightning strikes.

Many hurricanes were created in the oceanic hydro-sphere due to the high temperature and pressure differentials caused by the new radio-metric forces acting upon the Earth as a whole-energy system. The air, the waters and landmass were all experiencing temperature variants and pressure changes uncommon for ecological processes to proceed as regularly designed—waiting for Gravity-1 to return to normal range.

The whole Earth had moved unto its orbital space and beyond its allotted path, as if a whole-system earthquake took place underneath everyone's feet. There could be no escape: the planet behaved as one big rock to which all things and people were attached for maximum sensing of all vibrations, shocks and turbulences. The core heaved up with great emulsive flows that were constrained only by the massive plates that were so compressed towards Earth center of mass as to prevent magnum displacement of stratigraphic layers.

Many new springs and creeks were created due to underground waters being freed from masses of Earth encircling and restricting them. Some gushed out 30 feet into the air due to the great pressures that were released after compression terminated.

When the Earth re-settled into its pre-designed position, the plates re-adjusted themselves to cause pocket quakes localized in certain regions of the globe. Flooding took place along coastal towns and cities—many inhabitants, in anticipation of effects as warned from previous information broadcasts, had moved inland, which prevented the death of countless human beings on the Earth.

For a whole hour following the space-object's "grazing" Earth magnetosphere, atmospheric tempests, landmass tremors and oceanic tidal activities magnifying the shock, kept the whole Earth moving as a huge turbo-powered machine so boosted up as to require special time to process the saccades and counterbalancing gravi-pressure inputs to which it was subjected.

The increase in atmospheric pressure had "popped" many an ear drum as hospitals were swarmed with requests for medication and surgical repairs. A man reported not to have been able to "think within his head," but rather that he felt his thoughts externalized, forming outside of his body to then merge with his mind. He had also experienced a loss of hearing. He was summarily hospitalized. People who fell down had broken limbs or severe contusions requiring immediate attention.

The worse had passed away and now everyone was waiting for "sunshine after the rain."

Still, the death-toll was high; there was much crying, moaning, groaning and wailing as the living was morning the death of their loved ones. Vast quantities of property were lost—the lower the structure, the more probable its survival. In urban downtowns, as predicted by scientific analysis, many skyscrapers "bit the dust," as they were "jerked" from their foundations due to movements of the water table.

People will be busy clearing their neighborhoods of fallen debris, burying their dead, rebuilding their homes, and caring for the injured. Needless to say, in the United States, banks will be issuing reconstruction loans, including making other financial instruments available to the population for purposes related to repairing the vast damage that afflicted most regions of the country. Thankfully, industrial production, business enterprise, and commercial activity, having suffered temporary interruptions, had not been so severely affected as to preclude continued, sustained, restored operations.

*   *   *

15

ACH HEAD OF STATE was preparing words of constructive comfort, staunched determination and resilient
recovery, to be delivered to their people, respectively, in order to instill in them a spirit of cooperative unity that at the
same time encouraged individual initiative and resourceful-
ness. Simultaneously, international organizations touched
base in order to magnify common efforts at understanding cosmic events in a way that contributed to the highest
degree of livability for the human family as a whole. Not
knowing whether the sun's death was imminent was not a
reason for resigning to mere fate—the quest for knowledge
had to continue, however, now, to include the whole universe, bringing into factoring even interstellar variables that
situated themselves, outside of the purview of immediate
solar system configurations.

As was declared before, concerned scientists needed
a direct theory of solar entropy, a definite time span and a
constructive, specific saving solution for the human family.
To what extent that this solution will have to be publicized,
once adopted, was not known at that time, since qualitative decisions will have to be made regarding people and
resources.

Scientists again decided to meet in conference, an-
nouncing their decision to the mass media, but this time,
knowing in advance the precarious situation in the after-
math of the space-object collision with the Earth, without
any pressure upon anyone to attend. The meeting was still
an open one, and would be broadcast through the media.

Given current solar behavior, confreres came to an ed-
ucated surmise by common agreement that it was no longer
practicable to assume the sun's death to will have come in
about 5 billion years. The sun still had a lot of fuel. However,
due to interstellar variables that compelled consideration of
the sun as an "open system," whereas the previous estima-
tion was made solely on consideration of stand-alone star
types, they posited for the sake of clarity, planning objec-
tive, and projected resolution that, given the constellation
of sun-caused events, the sun will "die out" in about 800
years. That was the paradigmatic framework within which
they will have to prepare a saving resolution for the human
family.

They concluded that a theory of solar entropy that took
the interstellar medium and the far reaches of outer space
into account would also factor outer-galactic and interga-
lactic activity into solar system parameters in order to gauge
star behavior around planetary space—in the hope that the
intensities of solar gravitational and radio-magnetic field
variations could be reasonably or statistically measured in
terms of 'space curving' or 'space bending' activities. They
decided to apply the laws of thermodynamics in a way that
recognized Conservation, as well as acknowledging different degrees or kinds of Entropy.

They reasoned that Functional Operational Entropy
was occurring simultaneously with the march of eventual
Terminal Entropy, within a "time differential" that offered
opportunities for preparedness and readiness of action as
they engaged in building a saving plan for the human species. Does not a baby develop to grow into an infant, then
a teen-ager and then an adult, maturing, while at the same
time we know that we have mortality in our genes—that

eventually, human beings age in seniority and then eventually die? However, humans do not die immediately after birth—they have a lifespan during which to receive an education, to learn, to produce, to create and to live. Therefore, Qualitative Conservation or living life fully in the plenitude of good health and productive prosperity was occurring at the same time that the potentiality of accidental death or death from illness or disease "loomed" over human existence.

It was a fact that all things must reach final or Terminal Entropy, such as death. But developmental growth until maturity was counter-entropy operating. Since all things operated within ranges of functionality, then "terms" or "spans" or "durations" or "periods" of functional life engendered patterns of "negotiating" the "tugs of entropy." During the lifespan of each whole-energy system, each entity displayed a 'pattern of negotiation' with Terminal Entropy that took the form of corrective measures sustaining a temporal recalibration of Conservation, via a process called Functional Operational Entropy, during which they endeavored to achieve and maintain Qualitative Conservation.

Consequently, the sun, in order to maintain Qualitative Conservation, has been "negotiating" Terminal Entropy (star death) via the process of Functional Operational Entropy (radiating, shining, undergoing fusion, fission, convection etc . . .), by readjusting or recalibrating its nuclear energy processes within allowable ranges of plasma dynamics. And they commonly agreed that this process will take at least 800 years before the sun exhausted all its nuclear fuel to then undergo Terminal Entropy. That was their direct theory of solar entropy with a definite time span.

Consulting scientists reasoned it in this manner, "Since the sun is 'dying out,' 'piddling around' Earth orbit is not the 'saving solution' being sought. Only Earth-like, life planets can be inhabited. Space is made of emptiness and must be filled by matter in order to be habitable. Time is eternal—but we only have 800 years within which to resolve our problems. Space in infinite—but we have this solar system within which to live. Matter has mass, and mass presents gravitationally-sensitive parameters as explained by Relativity Theory—space around mass is 'curved' or 'bent.'"

"But the problem is that no other planet in this solar system is fit for human life. In addition, methods of space propulsion relying on gas ignition—Hydrogen and Oxygen—are desirable to the extent that they can be utilized for long-term space travel. Unfortunately, there are no other propulsion systems available. Propeller, rotor or blade systems as well as jet engines are for atmospheres only. Vacuum space has no atmosphere and thus necessitates combustive gas, rocket-propellant types of thrust fuel."

"What can be achieved in 800 years, given the contribution of all peoples and nations on the Earth to the collective human family saving endeavor? Could we encircle the planet with space stations—knowing what current astronauts go through to merely survive around Earth orbit?"

That was not considered a likely prospect since once the sun "dies out," so will the planet and all motions and forces that allow for orbital positioning, including gravity as an emergent property of revolution and rotation.

"How about a base on the Moon with plans for human habitation?" they asked. "Well, the Moon is a good place to construct a stop-gap measure for space exploration, such as an electricity-generating nuclear power plant. However, the

Moon itself has no atmosphere, no gravity, no plate tectonics, no magnetic field and no molten iron core with electroconductivity; and every thing we bring on the Moon will have to be taken out of the Earth system—thereby depleting Earth resources with no options of returning them for possibilities of recycling."

"Additionally, elements thought to exist on the Moon are just being 'skimmed off' the surface by solar plasma winds and are not intrinsic to the Moon's geological reservoir. The Moon has no inherent system of resource replenishment and all materials brought to it from the Earth will have to undergo 'Earth re-sourcing' before they can be tapped again for transplantation to the Moon. Some things like petroleum and metals, sweet water aquifers, and salt mines, are not like growing food every yearly season and harvesting every autumn."

"The Moon is not a planet—it's a rotating piece of rock, primarily basalt, with no determinant factors for 'terraforming,' or 'geo-conversion,' meaning that parameters for transforming it into a livable planet-type space-body, are totally absent. And furthermore, and more relevant to our complex dilemma is that the Moon is a natural satellite of the Earth—with the sun 'dying out,' there will be no gravitational effect on the Earth with which to 'keep a satellite,'—the Moon will not survive as a natural satellite. For it, too, will be thrown out of Earth orbital space and therefore, no tidal activity will continue, as there won't be any Earth for it to revolve around—which throws time-consuming and resource-gulping 'experimentation' out of the picture."

"Remember, we have little time for experimentation at this juncture, we need a constructive and appropriate saving solution with faithful assurance that it will work in sustaining the continuity of the human family into the future. In cosmic time, 800 years is not a long time, but for human productivity, it is indeed, long enough. As you know, the industrial revolution began in the 1840's and since then, it's been only 160 years, or so, and friends, just witness our great achievements! Did we not land on the Moon in AD 1969?"

"We believe we can make it in that time frame—in 800 years! And for comparison, we might consider that Christopher Columbus landed on this Continent in AD 1492 and this is the twenty-first century, in the year of our Lord 2009. So, from the Columbus landing, we're looking at a span of about 500 years for this Continent. But since we will be working in vacuum space where we have to transport every thing from Earth, then we have had the privilege of adding 300 more years!"

*   *   *

16

THE UNITED STATES OF America, Europe, Japan, Canada, India, Russia and China conferred together for a plan of action regarding "the saving solution" for the human family, each nation sending representatives, scientists, individuals and groups with various qualifications, credentials and expertise in all the fields of human knowledge whereby respective contributions might be factored into the elaboration of an agreed upon resolution.

The conference opened in Tokyo, Japan, at 5:00pm on Monday next, with worldwide broadcast of events and online internet connectivity accessible for questions and answers, as well as phone lines, and teleconference opportunities available to American and European counterparts who could not attend due to necessary assignments the functions of which had to be fulfilled.

The construction of space-stations or of a Moon-base was ruled out due to their impracticability as permanent habitable living spaces for human families in deep space, given that the sun and its gravitational solar system functionality will no longer exist for these structures to thrive. Another solution was necessary and all persons present were encouraged to come up with a suggestion, idea, insight, program, plan, or concept from which they could initiate a starting point for designing a new home for the human family in outer space.

All nations congregated in groups comprised of individuals from all participating countries which were structured to compete with each other in order to create the best concept solution for the situation at hand. A group led by an American, Lens Coralspring, was scheduled for its presentation.

The young man began his analytical *exposé*, "Look at the human body. It is constituted of 'near tubular,' or elongated members or parts akin to 'cylindrical modules' connected together so wonderfully by joints made up of small bones supported by ligaments, tendons and connective tissue. Look at a luxury train that imitated the human body, and the way its 'near-tubular' cylindrical carts or compartments are joined together in order to communicate with each other, while passengers safely and securely move from one to the other."

"We are proposing a system analogous to a long train constituted of near-tubular carts or compartments that will be taken to space one by one, one at a time, until a whole train of safely and securely 'jointed' compartments is constructed beginning at the farthest point within the solar system—that is, as far as our technologies and rocket fuel systems can take us. We cannot start building at the closest point to Earth or Earth orbit since this is where Matter-with-Mass, i.e., planets, already exists with its 'space-bending' or 'space-curving' forces; and since this is where they move to attract and repel each other, therefore, this is where the cataclysm will begin with solar death. And where gravitational fields operate, mass attracts mass, mass repels mass, mass 'bends' or 'curves' each other's space!"

"Where there is matter, there is electromagnetic mass; and where there is electromagnetic mass, there is a sphere of gravitational influence projected outwards from a center of mass, and from a center of field. The solar system has both a center of mass and a center of gravitational field. Therefore, we must position the "home train" in a location where only field determinants, if any, could affect its disposition in case of solar gravi-metric disturbances. Since it will be inert, as compared to bodies with electromagnetic mass,—it will have no magnetic field—then there is no reason to believe it will attract any gravitational processes akin to those which planets or moons are subjected to."

"But we are going to be filling up 'new space' so to speak, using train carts that are not planetary bodies having gravi-metric determinants that can attract destructive radio-gravitational connections with impending solar extinguishment. Consequently, we must make distance and the types of materials we utilize our allies in constructing a 'home train' for the human family."

"We will need the biggest rockets with the greatest amount of thrust in order to transport train compartments already built for functionality—their 'payload' areas will have had been built to sustain extended human activity, with respective appliances, machines, objects, and things needed for habitation, like a luxury ship, for example, in which the dining table is already bolted to the deck."

"These train carts will be designed to be built with light, sturdy and long-lasting materials such as plastic, titanium and aluminum composites, assembled with an inspection system for materials integration, artful quality, economy of space, excellence of craftsmanship, quality workmanship, refined technological adaptation, and highly technical instrumentation. For example, in terms of functionality, a luxury train has an engine compartment, a dining cart, a storage compartment, a sleeping compartment, a recreational compartment, a supply compartment etc . . . In the same vein, the 'home train' will have compartments designed to fulfill particular functions and provide specific services."

"This system is the safest, long-term method by which each successive cart or compartment will add living, storage, and resource fulfilling, encompassed spaces, within the tubular structure being constructed for human habitation. We have some accumulated knowledge and information from the space shuttle expeditions and space station habitation, wherein have lived many astronauts, of both genders, who will be too happy to inform us of their living experiences. Water, food, fuel, waste disposal, sleeping and living quarters, inner-breathing atmosphere suitability, gravity equivalent, clothing and washing, schooling and recreation, etc . . ., all these earthly necessities and conveniences that we enjoy here will also have to be configured, somehow, into these adjoining and successively transported compartments, that in the long term, will form a long train suspended in vacuum space, divided into different functions and adaptations to support human families living therein."

"The initial compartment will have to support the life of first astronaut-families leaving the Earth once and for all, with enough supplies and resources to last until the next compartment arrives with the new and additional astronaut-families, as they will be equipped with extra supplies, resources and materials. Remember, it is the human species or the human family itself we are preserving—there will not be any sun, Moon or Earth, any more, not in the livable forms that we are accustomed to."

"In 800 years from now, this area of the galaxy will be a de facto spatial wasteland, except to the extent we will have brought it 'back to life' by re-energizing selected lifeless planets in scientific ways that amplify the 'nitrogenization' and 'oxygenation' of their atmospheres, and the 'nitrification' of their geo-soil structures, to the proportions akin to Earth-types living planets."

"The Laws of Thermodynamics state that 'Energy is never created nor destroyed but always transformed,' meaning, there might be continuous transformations from one form to another, from one state to another. And we have the greatest determination, hope and faith, in bringing to fruition the 'transformations' we are now in the process of undertaking for the continuous protection of the human family and our posterity."

"Therefore, Terminal Entropy is generic to all individual universal system components and parts; but because there is also Conservation in the wholeness of that interdependent reality, thus there is also Continuum in its totality. Continuum Space-Time is a paramount reality because indeed, the sum is greater than its constituent parts."

"Integrated wholeness happens to exist in discontinuous quanta of mass-energy; however, its integrity is always sustained in the Qualitative Conservation of its inherent constitution, respective to its unified form. In short, some parts of the universe will have been 'dying' while other parts will be 'brought to life,' so to speak, in the very unfolding of that process. And human beings, with the help of God, will be the new life-givers, so to speak."

"The sun will have had by then undergone its 'death cycles' from a 'yellow dwarf' to a 'red giant' to a 'white dwarf'—and as we are hoping that it does not explode but rather "fades away" so as not to boil Earth oceans and evaporate atmospheric gases—once "sun death" occurs as a "fade away," we will have ample opportunity to work undisturbed by sudden bombardments of solar plasma ejections. However, we will have to account for random accretions of displaced radiation due to radioactive residues that will still be active in the solar material. But by then, as the 'home train' will be equipped with 'decompression chamber-carts,' so too will it be accoutered with protective radiation outfits, suits, chambers and compartments."

"With 'sun death,' in addition, no lightning will quicken the Earth hot iron core for molten liquidity. Earth oceans will freeze, its core will stop being molten and flowing, and its atmosphere will solidify, because only cold space will remain and will move in permanently as in a 'continuous winter,' hence, the remaining complex of which, engendering the vastest supply of raw materials for productive utilization."

"As concerns the amount of Earth mass that will remain, what happens to hot molten lava after a volcanic eruption? It becomes 'gray ash,' representing totally burnt magma, thus providing chemical compounds processed by the Earth and rain for top soil fertility. Given that the core will have solidified after 'sun death,' it is possible that great pressures will cause the plates to collapse with massive momentum weight upon the now-defunct ash-core, hence, causing a portion of the landmass to sink under the oceans, the plates having no more uplifting force from below. However, the majority or greater part of the landmass will have remained above ocean waters as everything will be frozen by cold outer space 'Kelvin temperatures.'"

"But for now, we have 800 years during which to not only begin designing, creating, and constructing, and lifting train-compartment rocket ships into space, but to also benefit from the foresight we'll develop in anticipating human Earth-needs in deep space as no other generation could have done. Space-Time is 'warped' only in the vicinity of matter with electromagnetic mass that 'bends' or 'curves' the spaces within their spheres of influence. Like all forces and processes in the universal matrix, they operate in ranges. Thus, it behooves our space engineers to locate and position the 'home train' beyond their gravi-magnetic or electro-radio-active reach."

"The laws of thermodynamics are immutable in their scientific applications—solar entropy cannot be denied. Only time-determinants need be factored in. These solar oscillations from coronal mass ejections to simulated apparent normal plasma-field projection activity are signs of unsuccessful Qualitative Conservation marked by events that do not correspond to past patterns of Functional Operational Entropy; they are "pangs of death"—imminent, 'star-death.' But as faithfully and fervently stated before, there is Conservation, and therefore, Continuum in universal wholeness."

"All scientists agree on these entropic processes but may differ on the time estimates for Terminal Entropy—but given the mass of the sun and its supplies of nuclear plasma fuel, we have scientifically deduced that this minimum time period of 800 years is sufficient for us, human beings, to make ourselves ready for space habitation in the irreversible advent of 'sun death'—as entropy is the fate and destiny of all things that functionally exist to perform in the universe."

"Rockets must be designed and engineered with the most advanced electronic guidance systems and with the most efficient fuel systems geared to discharge maximum thrust, hold and lift maximum payload, and travel maximum distances. Once the rocket leaves Earth atmosphere and orbital region, the rocket can be throttled at maximum velocity in order for it to travel to the longest distance from the Earth—and it will, with minimum effort since there won't be any gravity or atmosphere to travel through, just empty, void, vacuum space. And where it stops, if it ever does, will be where we begin 'inhabiting' in space. Retro-rockets would allow us to determine such a distance from the Earth as they would slow down or stop the spaceship-like compartments upon command."

"A question will arise whether to thrust or propel the compartment alone without the rocket in order to save rockets for future use. The answer is of course, no; additional rockets will be manufactured. Another question might arise such as why not send remote controlled space carts or compartments without risking the lives of human passengers. The answer is simple—there will be vital work to be done for the survival and living viability of future generations, and only human beings have the intelligence and creativity to perform them."

"The rocket engine must always accompany the compartment in order that it has a way of 'navigating' through space. Retro-rockets will allow it to stop and position itself in order to 'couple' with the next incoming compartment. A certain distance from the Earth will have to be chosen or compartment-rockets will continue to travel into space indefinitely."

"The space shuttle had external rocket boosters as well as internal engines to assist lift and maneuverability. Likewise, the compartments will be equipped with internal engines and retro-rockets located in their rear and below the connecting floor levels, in order to assist in lift and maneuvering in space. These compartments will have external rocket boosters as well, to lift them from Earth gravity and into Earth orbit, and likewise, they will be jettisoned, and those will be re-usable."

"When constructed of light but durable composite materials, compartments will be easily lifted into space to then be maneuvered into positioning for connection with the previous ones already established and located in the region of space selected for habitation and social arrangement."

"Just picture and envision an habitable train of continuously linked and connected living compartments and carts, stretching from the farthest point from Earth within the solar system, and in a direct line of communication with Earth systems of space transportation. At this time, it has been surmised that, though an endlessly rich array of configurative applications as to structural arrangement and compositional positioning could be approached, the 'home train' could best be structured in a straight line configuration in order to afford safe and secure compartment connections from one to the other."

'But as compartments continue to accumulate into a great agglomeration of connected carts, a circle structure configured like 'the old wagon trains' would have to be considered for freedom of movement and more efficient internal operations. By continuously re-adjusting its angular momentum, we could avoid the trappings of a limiting circumference, the logistics of which would allow for endless additions of carts or compartments. Compartments connect at their ends and not at their sides—though the respective fittings and openings or 'hatches' for side connections will have been included into their intrinsic design and manufacture. Further down the road into the 'home train' construction, 'side-jointed' additional carts will be shipped from the Earth to join all 'home train' compartments—in a way analogous to latitude lines that 'join' the Earth all along its peripheral circumference,—for ease and freedom of movement within the 'home train's' internal surface area."

"But, in the beginning, until the circumference of connected compartments is great enough, straight-line connectivity is best practiced for it will make it easier to enlarge the circle of 'wagon train' circumference as more carts or compartments are brought in for addition. Friends, we have 800 years to put it all together."

"But this is only the first part of the constructive saving solution for the human family. At the same time that rocket-lifted compartments are being transported into space, other sources of materials and supplies must be sought out—in case of 'sun death' or supplies are exhausted on the Earth, whichever comes first. Other planets must be taken into consideration, as 'source planets.' And these developments must take place before 'sun death'. We must factor these possibilities into the space-building equation. For, an equation has two sides, each the equivalent of the other, respective to the conversion factors or 'constants' being applied—the Conservation side and the Entropy side, as connected by the sign of equivalence, the equal (=) sign, with consideration of anticipated but yet-to-be discovered variables."

"Therefore, during the 800 years preceding 'star death,' we must also endeavor to re-engineer an earth-like planetary system. It is not probable that other planets will continue to revolve around the sun and rotate upon themselves in the absence of gravitational solar activity, but, for operational and logistical purposes, it must be assumed during the whole period of time preceding 'sun death,' until we can prepare and make ourselves ready to re-provide these other planets with gravitational inputs to process into a new system."

"Due to the necessity of needful materials and resources, and the possibility of "vivification," a process whereby a formerly "dead planet" could be "brought to life," we must factor 'materials mining' operations from 'geo-transformed' planets into the space-habitability formula."

"Building the 'home train' will require vast amounts of materials and resources; so will planetary 'vivification' or 'geo-transformation'. There will be need for a continuous source of raw materials regardless of when the sun finally 'fades away'."

"For, this question is most urgent on our minds, once the sun 'dies out' and the Earth is no more, where shall we obtain materials and resources to live and to work with in Space? Will we have to transplant soil and water continuously, while the Earth is still here in order to grow plants and food stuffs also in space, by engineering green-house compartments as well? Will we need 'livestock-raising' compartments? Or can we continue to ship food from Earth 'while the going is good' and the sun still with us? For example, food can be canned in composite plastic "bags" or dried and preserved in plastic containers—the lighter method being the best."

"To our advantage, that is, before the 800 years expire, we will have at our disposal the Earth, the 'home train' under construction, and other planets or another planet chosen for 'geo-conversion'. The problem is with the timing of 'sun death'."

"Therefore, due to this principle of uncertainty, we have been thinking of a more long term alternative to these processes, as we must prioritize actionable operations in a way that anticipates developments, such as the case of 'star death' before completion of 'vivification' is finished."

"But what will happen when Earth oceans and the atmosphere freeze, and the core solidifies? Yes, resources such as soil and frozen water might still be available as raw materials; however, how would we avail ourselves of them? Wherefrom would we obtain materials to build "water tankers" or "soil transplantation tankers" with rocket engines of great thrust and propulsion power? Could another planet's atmosphere be transformed into hydro-forming constituents of salted or sweet water? How tedious, and time consuming, and resource depleting, would it be as compared to 'vivifying' the whole planet instead?"

"Remember, we will have little time for 'experimentation,' but rather, will have been building in space while having the Earth as a primary source for materials, resources and supplies—while the sun was still there and the Earth, still, a living planet. However, once these conditions no longer prevail, all we will have had left will be the train compartments joined together in a great circle, and the degree to which we had built the 'home train community' as a place of technological advancement and creative productivity."

"Now, you see, my friends, the great necessity of 'home train' economic development and social productivity that will have by then duplicated Earth levels of achievement! Are we not 'exporting' the utmost best of everything into outer space? Yes, we are, and therefore, we must contemplate that in the span of 300 years, our 'home train' will be a 'booming,' 'teeming,' and 'titivating' society equipped with all the wherewithal of modern life—in outer space."

"Thus, must come the second part of our saving solution—the Earth does not have an inexhaustible supply of resources, on the one hand, and the sun is going to 'die out,' on the other. In other words, we must replace both or duplicate both the sun and the Earth, all at the same time that the 'home train' is being constructed. And this is where the ingenuity of the human family and our capacity to create great things come into full, wonderful appreciation."

"On the one hand, we must duplicate the sun—by creating an artificial sun; and on the other hand, we must transform one of the lifeless planets into a life-giving, Earth-type planet on which human beings can live. For, we will still need equipment, spaceships, rockets with great thrust, advanced technologies, sophisticated electronics and competent machinery, as they will all be necessary for planetary 'geo-forming' and 'resource mining,' even in the event of precipitated 'star death.'"

"Some necessities like food can be replenished every year via the planting season and livestock farming can supply us with meat on a steady basis provided animal food supplies are also replenished. However, once this sun 'dies out,' we will need another one; and also, another space-body or planet will have to be 'cultivated for life' or 'vivified' for human habitation, especially for land cultivation and livestock farming. We believe that both can be achieved within the span of 800 years, including continuing the construction of the 'home train.' Let us explain."

"So we have these three undertakings going on at the same time: building the 'home train,' creating an artificial sun, and 'vivifying' a lifeless planet for human habitation."

"At the same time that train compartments are being lifted into space and assembled in the selected space region of habitability, enough human families will have had established themselves, so that, in the span of say 50 to 100 years, a pattern of space productivity, Earth-resource replenishment, and social development will have had emerged so as to instruct us on newly arising needs from those that have already been successfully fulfilled."

"In the instance of creating an artificial sun, we already know that the sun's nuclear plasma is 'suspended' in vacuum space, within a strong magnetic field. We shall have already been in space and developing methods of 'extra-vehicular activities' that extend beyond the activities that take place within the 'home train.' We have had the benefits of the knowledge of fusion and fission processes on the Earth because of nuclear research that led to the creation of the atomic bomb and the Hydrogen bomb. Such processes are not compatible with Earth living environment. However, deep space is already bathed in nuclear radiation and timing the creation of an artificial sun with the impending 'death' of the solar system sun will prevent any magneto "interfero-metrics" or "electro-arc" that could have been engendered by trying to invent or manufacture an artificial sun while this one still exists—we are assuming that the critical mass, size and radiating power of the artificial sun, which we will be able to control, will not be sufficient to create gravi-in-terference with the current sun. We will thus proceed as circumstances warrant. If 'interferometrics' are present, the artificial sun or self-contained nuclear fusion reaction chamber will have been prepared and made ready 'to fire' as soon as the present sun 'fades away.'"

"While the solar system sun is here, we will still need its light in order to work in dark vacuum space. Or by that time, we will have had manufactured plenty of artificial lighting, including the production of long-life batteries and the design of long-life solar cells that collect solar radiation for storage and for conversion into electrical current that can power tools and be utilized for other purposes."

"As you understand by now, activities aiming at 'vivification' of the lifeless planet will have to be on-going so as to re-engineer the development of a 'greater home' for human beings, while Earth is still with us."

"During such a time, scientists will have had already researched the best materials suited for nuclear chain reactions and the best methods by which to generate a strong magnetic field in space within the generated forces of which to 'suspend' the initially ignited, fused nuclear plasma. The strongest magnetic field is needed to suspend or contain the nuclear plasma. Nuclear engineers will also have to determine the extent to which it can be 'fed' in order to control its critical mass and size per emergent gravitational effects upon any neighboring planets and per solar-effect parametrics."

"For example, the planet Venus is almost similar to Earth in size, gravity, bulk mass, and geo-composition; Venus, has a mass closer to Earth mass, more than any other planet, possibly possessing an iron core like the Earth, and gravity that is almost the equivalent of that on the Earth. However, Venus is already the second-closest planet to the sun and receives so much solar radiation that it would be impracticable for it to receive any more. To the contrary, either it should receive less solar radiation or it should develop an atmosphere akin to Earth atmosphere that would allow it to retain a magnetic field, hence, initiating its ability to cool itself instead of absorbing and reflecting so much solar heat."

"Thus, where the artificial sun is assembled becomes a matter for crucial consideration and utmost attention. It has to be at a safe distance from the 'home train' and also far enough from the current sun, and from a specifically targeted planet like Venus, the life-transforming parameters of which, will need to be controlled in order to engineer 'vivification processes.'"

"And at the same time that the artificial sun is being built, we will have to decide which planet to 'vivify' or 'transform' into an Earth-type planet. As mentioned above, Venus is the most promising prospect. Jupiter, Saturn, Uranus and Neptune are not only too far, but they are also classified as 'gas nebula,' and having no 'silicate geo-structures,' analogous to rocks on the Earth. In contrast, the so-called "terrestrial planets," like Venus, Mars and Mercury have 'rocky' geo-compositions."

"Of the three terrestrial planets, Venus appears to be the most promising for 'geo-transformation.' Scientists, specifically geo-physicists, earth sciences experts, geo-chemist and bio-physiologists, to name a few, will have to devise methods of deciphering the ecological code or mechanism via which to convert a Venusian atmosphere of mostly carbon dioxide into one of mostly Nitrogen, like the Earth atmosphere."

At this time, the young man, Lens Coralspring, motioned to his assistants to start distributing a pamphlet-like material constituted of stapled sheets of paper in which comparative data are given for evaluating Earth-Venus compatibility as to "vivification potential."

He continued, "Oxygen is present but trapped in the form of Carbon Dioxide, close to 96 percent of it on Venus; and Nitrogen is also present, 4 percent of it on Venus. There must be researched and discovered catalytic geo-chemical reactions that could cause changes in atmospheric composition, structure and content on Venus, from 96 percent Carbon Dioxide and 4 percent Nitrogen, to close to 80 percent Nitrogen and 16 percent Oxygen, with Hydrogen,

Neon, Carbon, and other gases constituting the remaining 4 percent."

"Do we not have reaction chambers on Earth wherein temperature and pressure are controlled as range variables in order to determine molecular bonding for new compounds and new chemical formations? Why could we not utilize these very processes in order to discover those catalytic compounds that would trigger reactive mechanisms on Venus which would then engender the organic molecular transformations in its dense carbon dioxide atmosphere?"

"Though closer to the sun than planet Earth, Venus presents the closest metrics in variable form and allowable range that would be amenable to 'geo-conversion.' Per atmospheric composition and geological structure and content, in addition to the need for an electro-conductive core that can sustain a magnetic field, Venus could benefit from a new geo-chemistry that would present new inputs for it to process in order to obtain life-supporting outputs."

"The Input-Process-Output mechanism now operating on Venus is digesting variables and factors that are co-determinants of its apparent uninhabitable conditions, probably due to its proximal distance to the sun and its absorption of great amounts of radiation, heat and light, without the capacity to 'filter,' 'buffer' or 'extenuate' their gravi-metric thermal effects, except by reflecting them. Yet it has a dense atmosphere, but of the incorrect constitution for human habitability. Thus, scientific ingenuity must be coupled with the understanding of the catalytic processes that engender a breathable atmosphere on the Earth, its possession of a magnetic field and an electro-conductive molten iron magma core, as well as an atmosphere composed of function-specific layers, and a liquid oceanic hydrosphere, all of which, being indispensable knowledge for 'geo-conversion' of Venus."

*   *   *

17

THE WORLD OVER, IN the news media, schools, universities and households, widespread discussions and elaborations by scientists and other people alike regarding space habitation and "planetary vivification" propositions abounded with a bountiful and rich diversity of ideas and opinions. However, to most people, human beings were having a "reality-check" on the fragility of human life in the universe and the limitations on scientifically determined natural laws imposed by Relativity operations that govern Earth processes and universal events due to the unstoppable march of the Laws of Thermodynamics, especially the law of entropy.

On network, and cable television, and radio, in the United States, an open forum began, linking them all, whereby station personnel, appearing guests and interested callers could talk, converse and communicate all at once. Multi-media teleconferencing was the most ubiquitous novelty taking place in the country. This helped disseminate information at the same time eliciting participatory responses from the population in a way that contributed to their sense of belonging, well-being and societal security in a world where the very physical universe itself was in upheaval. Somehow, there was a common unspoken understanding that though they could not control solar activity, they could have self-determinant control over their own mental activities that warded off negative thinking or destructive approaches to problem solving.

The "software of human thinking" did not have to succumb to the chaos apparently emerging within the "hardware of universal reality."

A woman whose first name was Barbara began the exchange with a succinct summary of the apparent facts that controlled our fragile existence on the Earth and in the solar system.

"Here it is folks. The sun is 'dying,' the Earth and the Moon will disappear as we know them to have existed as life-support systems—they might exist in some form of lifeless matter-mass or other, but no longer suitable for human living and habitability. In addition, while we're waiting for the sun to complete its now-accelerated 'death cycle' to a so-called 'white dwarf,' Earth resources, materials, and supplies are not limitless, they can be exhausted. And as far as we know, as the Lord has already willed it, only food can be re-planted, re-grown and re-harvested from season to season and year to year. Trees may take 40 to 80 years to mature, even if replanted. Everything else from oil to coal, from wood to rocks, all, have finite quantities even as we are processing them for human utility on the Earth, with the law of entropy not too far behind in restricting the quality, quantity, and period during which we can utilize them. We have begun to recycle certain things, like plastic, glass, paper, spent oils, and metals, but not all materials are recyclable and oftentimes, it takes more resources and energy to recycle them rather than to begin from scratch, such as in mining metals and digging for oil."

Another person, a man named Gerald, continued, "And when we are engaged in traveling to outer space, what must happen? We literally have to take the Earth with us and bring it with us in our 'back packs' so to speak. And every thing we need for space exploration, other expeditions and projects such as for the space shuttle or the space station, every thing we need in outer space must be deducted from the extant supplies on Earth, as you have so eloquently described, many of which are not 'replenish-able.' We keep talking about oil wells drying up—that's true. This can happen. And envision this—so many of our chemicals and materials are petroleum-based, from plastic to antifreeze."

"Correct," said Debbie, "You have touched on the major sources of materials used by human beings since the beginning of Creation. Rocks, trees, metals and oil appear to make up the list of ingredients of which civilized living is made. Even metals can be exhausted. I mean, look at your vehicle—metals, paper-goods, petroleum-based products like plastic, and rubber. That's it. Rubber is a plant and can be seasonably harvested. But metals and petroleum, having finite deposits under the Earth crust will eventually be exhausted—hopefully, not before the 800 years have expired so that whatever plans scientists and other concerned parties may come up with might be realized for the benefit of posterity and continuing generations. God help us all!"

A caller, Martin, added, "So we have about 800 years to create a new home for humans in space and then try to transform Venus into a living, habitable planet. All initial production and manufactures will have to be performed on Earth and from Earth materials. If we don't have other sources for materials, it's true, we could exhaust Earth resources. But given the circumstances, we have no other alternative, at least for the foreseeable future. From the events that have occurred in the past two weeks, loss of human lives, the destruction of property, the re-structuring of geology, flooding, fallen buildings and structures, I mean, folks, this is serious stuff—a catastrophe of immeasurable proportions! Life is so precious and yet so precarious on the Earth at this time. This, I believe is a 'wake-up call.' And the sun, the main problem in this equation while at the same time being the life-giving star for the system as a whole—is misbehaving, so to speak. It's like 'being caught between a rock and a hard place.' In this situation, my friends, only God in his merciful might and unconditional love—He is our refuge and counselor, He is our deliverer and salvation."

A woman at the television network, Marla, joined in, "We will indeed need the Lord, for, from what I have heard, the atmosphere on Venus has no transparency due to sulfuric acid clouds reflecting light back into space. It's the kind of acid that's in our vehicles – toxic, lethal and non-breathable. This sulfuric acid opacity might be caused by solar radiation in coordination with planetary pressure and temperature due to the fact that the atmosphere on Venus is more than 96 percent carbon dioxide with voluminous amounts of Sulfur floating in its clouds; hence, its great density for trapping heat for chemical reactions under immense atmospheric pressures, such as in reactions that engender cloud formations of SO_2 and H_2SO_4. Nitrogen content, being only at 4 percent, is not of a proportion great enough to effect atmospheric stability in terms of pressure and temperature, and for purposes of 'nitrification.' Sulfur in the atmosphere might have been propagated by internal geo-thermal chemical activities that have ceased or are on-going, analogous to earthly volcanic eruptions."

"But the main facts to remember, I believe, are that the gases of which a livable atmosphere is constituted such as Oxygen, Nitrogen, Hydrogen and Carbon, for example, are also present on Venus, but in different inorganic forms or compositions, such as in carbon dioxide, sulfuric acid, sulfur dioxide, and elemental Nitrogen—as CO_2, H_2SO_4, SO_2 and N. Given how short our remaining time seems to be in comparison with Eternity, it behooves us then to fashion the greatest possible constructive and beneficent overall prosperity for ourselves and our posterity—scientists have to discover those 'catalytic compounds' for Venusian atmospheric reactions towards an Earth-type living planet."

"Well said," continued Timothy, "I believe that scientists on the Earth ought to be able to ferret out and discover the necessary chemical reaction steps as to needed catalytic factors that would trigger transformation or energy conversion on Venus, from carbon dioxide, sulfuric acid, sulfur dioxide, and elemental Nitrogen, to atmospheric Nitrogen, Oxygen, Hydrogen, Carbon, and eventually, to water vapor or condensation. For, water is life. For, water will be necessary for the process of "nitrification"—whereby the formation of nitrates can be introduced via nitrogen fixing soil bacteria."

A caller, Russell, enjoined, "Yes, however, on the Earth, these catalytic steps are intrinsic to geo-atmospheric processes and eco-biosphere events, and have already been activated within the Input-Process-Output ecological mechanism, and thus will have to be re-discovered, so to speak, by humans who inhabit this living planet."

"'Energy is never created nor destroyed but always transformed.' Geo-chemists, bio-physicists and microbiologists, among others, will have to cooperate in research explorations, by starting with these elements, compounds or chemicals, such as carbon dioxide, sulfur dioxide, sulfuric acid and Nitrogen, in attempts to synthesize the processes via which energy conversion can be constructively and beneficially induced in the laboratory, by the introduction of catalytic converters, in order to arrive at life-supporting elements, chemicals, compounds and molecules."

"I believe it can be done, but they will have to re-trace the steps of natural chemical transformation patterns already operating on the Earth, so to speak, in order to duplicate them, first in the laboratory, and then, in a lifeless environment such as on Venus that already presents opportunities for such transformations to occur. Furthermore, there are advantages to 'terra forming' or 'geo-transforming' experimentations on Venus. For, due to Relativity Theory applications, we know that the property of equivalency has established that, a change in parameters or variables on one side of the equation will necessitate a corresponding change in parameters or variables, on the other side of the equation, as co-determinant factors are 'relative' vis-a-vis the common standard, the velocity of light, which cannot be breached by any event, process or object in the universe; meaning that changes and transformations brought about by human intervention on Venus might trigger Relativity modifications such as in temperature, pressure, magnetic field, core status, geologic structure, tectonic activity and the like etc . . ., amenable to human livability, and therefore, to human habitability."

"We thank every one for their participation and this concludes our open forum for this evening," said the television network announcer. "Join us again tomorrow for a continuation on scientific developments and ideas of our age."

*   *   *

18

THAT THURSDAY NIGHT, THE Moon's brightness was attenuated by unusually colored clouds that imparted to it a "reddish taint" bordered by a thin "yellowish bubble" as it struggled to shine behind them. The Moon displayed a spherical symmetry evoking its temperate locomotion as Earth's only natural satellite whose effects thereupon were commonly agreed upon to be mostly beneficent. However, because the Moon's center of mass intersected Earth center of mass at about 1100 miles beneath Earth's surface, at a point called the "barycentre," or where centers of mass meet, the tidal expanse of the Moon's momentum gravity effects manifested themselves on the Earth as "ocean tides" or "Moon tides," due to pressure-temperature differentials that worked in conjunction with Earth-caused oceanic motion parameters which also contributed to wave and current activities.

That night there was a full Moon. The weather was warm and star light could be seen through various densities of cloudiness in the skies above. The temperature was however a little humid but tolerable, at about 72 degrees Fahrenheit. Winds began to pick up at about 8:00pm whereby trees started to sway from side to side and branches toggled up and down. Leaves fluttered as atmospheric molecules made their way through the foliage with fluid strength that plucked out individual leaves once in a while.

"Mom," asked Annette, "How come the Moon is moving while we're standing here?" Annette, about eleven years old, was standing on the porch of her parents' house, while the whole family was enjoying the autumn evening by taking a break outside from the daily chores they attended to along the way. Annette's parents were Tammy and Ronald Zinlinnox. Tammy and Ronald looked up at the Moon and Tammy exclaimed, "My goodness, it is true Ron. The Moon is racing around the Earth with such a speed as we have never seen before. Usually, it is just there. And one might notice its motion around its Earth orbit, once in a while; but never like this. What do you make of it?" she asked.

"Well," said Ronald, "I am sure, they, I mean the people responsible for watching and observing the universe already know about it and it would serve no purpose to call. But let's check with the news and see if they're up and up on that."

Tammy and Annette followed Ronald inside the house. They turned on the television set on a news channel in order to listen to scientific developments, if any. They were "right on target." NASA and other space agencies had not missed the event and cable news networks were beaming commentaries and explanations all over the Earth.

"Good job, Annette," said Tammy.

"Yes, Annette, you would make a good scientist," enjoined Ronald.

The broadcast reported, "Though the Moon is geosynchronous—meaning that it rotates on its axis during the same period of time that it orbits the Earth itself—it has been determined that the unusual velocity with which the Moon is orbiting the Earth was due to solar gravitational activity and perhaps to tidal effects by other planets, such as Jupiter, Saturn and Mercury, in reaction to solar pressure-force differentials straining mass momentum factors out of galactic inputs impacting the solar magnetosphere, the causes of which are unknown at this time. Earth revolution around the sun has shown no changes while however its rotation rate has slightly increased. Please do not change your clocks—as of this time, it is assumed that it will still take 24 hours for the Earth to complete one day or one complete rotational cycle upon its own axis, 'give or take' a few seconds."

"The Moon's increased velocity was noticeable while the Earth's rotational velocity rate increase may be less observable due to the mass differentials between the Earth and the Moon—the Moon is about one eightieth of Earth mass."

"Scientists are not worried about the Moon 'flying out of orbit,' so to speak, due to the fact that it is 'locked' into its path in accordance with Relativity variables that are not so easily dislodged—Earth mass is much greater and therefore will exert its gravitational pull upon the Moon in order to stabilize it eventually. However, there might be effects on hydrosphere activity and ocean currents as affected in terms of water temperature, pressure, and round-earth circulation due to new multi-vectored motions, thereby influencing the development of yet-to-be-forecast weather patterns."

"In addition," the reporter continued, "Planet Mercury has displayed a 'double shift' not just at perihelion or at its closest distance from the sun, but also at aphelion or at its most distant proximity to the sun, or at the other end of its orbital plane. This event is very uncommon for Mercury whose perihelion shift is well known and predictable."

"Jupiter, the planet with the greatest mass in the solar system, next only to the sun, underwent 'wobbling gyrations' as it began to absorb, for a change, much more energy than it received from the sun. Usually, Jupiter reflects outwards much more energy than it absorbs. It appears that this process has been temporarily reversed due to internal emission-absorption events within its gas nebula where highly pressurized electro-conductive 'running flashes' have been observed."

"It has been surmised that all these occurrences are due to gravi-electromagnetic forces originating at the outer edge of the heliosphere where solar plasma has interacted with galactic spectral mass emissions from nearby star systems and supernovae. It is now definitively undoubted that the sun in this solar system is also receiving inputs from the interstellar medium, reflective of activities that have motion-sensitive portent for the solar system as a whole."

Suddenly, the whole Earth quaked as if having received a big shock as the Moon slowed down and began to return to its regular orbital velocity. Some old bridges collapsed; some roads caved-in. Some drivers, trapped underneath the rubble were being rescued by other drivers, neighbors, fire departments, state police units, city, and county ambulances.

People have shown great forbearance under these circumstances, knowing fully well that solar activities are beyond the control of any earthly authority—there is no one else to blame. Therefore, they did the best they could in assisting one another in alleviating pains and assuaging emotional grief and psychological trauma.

As if to add to human and earthly misfortunes, though it is night in the United States or this side of the planet, it is however day light in the rest of the world or the other side of the planet—for, effects were being experienced by some in daylight and by others during night time. Those on the night side of the Earth will have their "moon light" taken away "right from under their very eyes." The Moon attempting to regain its angular momentum for regular orbital positioning and normal velocity had to interface with Earth increased rotational velocity in order to restore stability to its moving mass—in short, the Earth too had to "slow down" at the same time, and at a rate that would sustain synchronized re-alignment and re-positioning for faithfulness to the Moon's normal orbital pathway.

As the Earth's rotational velocity underwent these re-calibrations, synchronization was achieved at the price of an unscheduled and un-forecast lunar eclipse.

No one knew what was happening within the sun. But somehow, the Moon, the Earth, and the sun were all aligned in a straight line, but with the Earth between the sun and the Moon, thereby engineering the temporal disappearance of the Moon—a lunar eclipse had "dawned" upon humanity as the skies got "darker."

"We apologize for the momentary lapse in broadcasting, ladies and gentlemen," said the reporter, "as you know, the whole Earth had just trembled as the Moon attempted to regain its regular angular momentum for orbital re-positioning. There is a lunar eclipse 'to boot.'"

Scientists engaged in systems analysis encouraged everyone to "remember exactly that—that we live in a solar system where planets, as well as the sun, possess spheres of influence, the activities of which depending upon their masses and their own attempts at 'negotiating' solar entropy inputs, and galactic inputs from beyond the interstellar medium. The Moon does not have enough mass to have caused it to 'jerk' the whole planet however; there could have been inputs from other planets, the sun and the outer reaches of the Milky Way Galaxy."

"As you have just learned, Jupiter and Mercury are also affected in ways not conducive to system stability. All planets revolve around the sun in almost circular patterns as on a three-dimensional disk-like plane; the plane is rather an ellipse akin to a 'flat ellipsoid' with minor eccentricities around the system's center of mass; and some planetary orbital ellipses are greater than others around the sun—with perihelion and aphelion, based upon mass, velocity, and distance from the sun."

"That system is held together by tremendously powerful gravitational forces as configured by motions of the Milky Way Galaxy that keep the sun in its proper position and by gravi-metric plasma dynamics that drive the solar inferno in its discharges of electromagnetic radiation which activate Earth for life-supporting motions of revolution and rotation."

"The Moon is 'caught in between' as the Earth itself possesses gravi-metric binding energies from its magnetic field and momentum mass, endowing it with the capacity to attract and retain natural satellites—the patterns of which is duplicated in the positioning of artificial satellites around Earth orbit. The Moon exhibits resistance to motion due to its momentum mass, thus, the demonstration of small variations in its orbital path around the Earth. The Moon is understood to affect seasonal weather patterns as well as hydrosphere-centered oceanic currents and air flow that contribute to storm systems and other geo-sphere processes and events."

"The sun also rotates on its axis, albeit very slowly due to its momentum electromagnetic mass, which, with other plasma convection factors, produce a nuclear-driven electricity generating dynamo. The solar dynamo transforms nuclear plasma mechanics into gravi-metric electro-magnetic dynamics from which Earth obtains its life-support systems, from its magnetic field to its electrified, lightning and rain-producing atmosphere, to its electro-conductive hot iron molten magma core which, in conjunction with the oceans, keeps the Earth within the range of livable temperatures."

"As you know, deep Space is extremely cold, even with a star bombarding it with continuously hot radiation. Thus, it is via seasonal adjustments, that Earth core, oceans and atmosphere work together to sustain temperature equilibrium. Both the Earth and the Moon receive inputs from the sun, which itself is not immune from galactic pressure-force inputs of momentum-driven radiation mass."

"We have just learned, ladies and gentlemen, that what might have caused the lunar eclipse was due to events happening within the sun as the sun was positioned right in front of an extremely dense star group. Due to its having been bombarded by outer-galactic gravi-radio plasma-tidal inputs from a group of bright stars called 'the Hyades cluster,' the sun's magnetic field had 'jerked' its compressive mass so as to alter its axial rotational angle, hence causing catastrophic transformations in its spectral mass radio-gravitational emissions with grievous disrupting effects for Earth-Moon synchronicity."

"These complex momentum forces upon the Earth and the Moon caused them to accelerate as the sun itself underwent a sharp deflection in its rotational axis angle, emitting great bursts of energy, during which its rotational velocity slightly increased due to its massive bulk responding to the 'Hyades star cluster' bombardments. The Moon's acceleration rate was greater due to its lesser mass."

The network then ended its broadcast with these words, "We have just received word that Mercury has resumed apparent normalcy in that it displays shift only at the perihelion—at its closest distance from the Sun. We thank you for listening and watching. We now resume our regular programming. Stay tuned for further treatment of these scientific earth-shaking events, no pun intended."

*   *   *

19

THE LAST TIME THE sun had a polarity change was in 2001. The polarity change cycle is every 11 years, concurrent with the culmination of its sunspot activity, the next one being due in 2012. This happens as the magnetic field undergoes an exchange in poles—the north becomes south and the south becomes north. It is a transitional phase marking the solar maximum—when sunspot activity is at its summit or climaxing peak.

Sunspots appear to be multi-polar pluri-potent "pocket magnetic fields" competing with the external or dipolar solar magnetic field in a way that prevents heat exchanges within the sunspot region, hence, demonstrating cooling by their "darker appearance." The "darker hues" could also be indicative of "carbonized materials" accumulating due to periods of "plasma short-circuits" within the solar core, expelled as "spent fuel," so to speak, to the photosphere due to their incapacity to combust as fuel reactants in the nuclear chain, until they are re-cycled again by the convection zone.

The "pocket magnetic fields" flow with cumulative opposite charges towards each pole, and hence, causing the weak external field polarity to yield to the cumulative oppositely charged polarity, thus causing a "polarity change." The sun's magnetic field is a huge envelope that powers the extensive heliosphere—covering a distance of about 75 astronomical units, one astronomical unit being the distance between the Earth and the sun or 93 million miles. Polarity changes are often accompanied by "solar winds" constituted of plasma particles and stellar rays that have extreme repercussions for planets in their wake.

Galactic and solar observations have been on-going in order to detect any unforeseen or unpredictable forms of radio-magnetic activity by novae, stars and supernovae, beyond the galactic regions, just in case they might have ramifications within this solar system for which there could be preparative steps towards damage control. It has been reported that novae at the extreme interface between the Milky Way and Andromeda galaxies where tidal effects are possible, have been showing increased spectral intensities in the X-ray range whereby the suddenly generated brighter luminosities had corresponded to huge explosions extending several astronomical units of distance, the exact relative distances of which have not yet been conclusive.

Consequences of these observed explosions for the solar system and the sun have not been assessed, except that astrophysicists have informed all channels in the information dissemination business and in universal discourse activities, to remain available for announcements to the public at large.

Scientists have hinted that, were the sun's activities to be symptomatic of greater internal, yet unknown plasma thermo-radiative core and atomic energy field momentum dynamics, it might undergo an "early polarity change" that would be accented by extreme radio-magnetic turbulences and perturbations, disrupting its hydrostatic equilibrium— the dynamic equilibrium between its thermal radiation energy and contractive gravitational compression force, which is countered by outwardly vectored controlled explosive forces subsumed in its convective flux dynamics during which great amounts of energy are released.

NASA, the Marshall Space Flight Center's solar research group, the Canadian Space Agency, European Space Agency,

Solar and Heliospheric Obervatory (SOHO), Big Bear Solar Observatory, along with the Russian Academy of Sciences' Space Research Institute, Japan Aerospace Exploration Agency, and many others in the world, have been analyzing these data in attempts to deduct logical inferences regarding the ways in which this solar system might be affected. It is a daunting task since these universal events and solar processes are beyond the reach of any human control, while the scope of discovery has been circumscribed by the limits of technological apparati.

Additional probes have been summarily dispatched, during 'windows of opportunity' as afforded by apparent 'solar tranquility,' to perform outside of Earth orbit and towards the center of the Milky Way galaxy in order that closer observations could be made from which they could broadcast and transmit cosmic and stellar data recordings to these various space organizations.

"What happened to the Hydrogen spectral lines," asked Stewart Tongurbell at SOHO.

"Check the other two spectroscopes," replied Colbert Arbielfung.

At 7:00am on that Wednesday, all spectroheliographs displayed the same data—no spectral electromagnetic signature for Hydrogen was being displayed at that time. And all machines had been previously checked and calibrated for accuracy and efficient functioning. To Stewart and Colbert, it appeared that something was taking place to which everyone should be alerted.

"Expect the worse to happen, my friend," said Colbert. "By the way, what's going on with spectral signature for Helium?"

"Wouldn't you know it? Look at this!" exclaimed Stewart.

The spectrograph for Helium was larger in size and greater in intensity and magnitude. All agencies, organizations and personnel were communicating in order to verify these events were really occurring and were not due to technological errors. All reported the same recordings all over the Earth. Therefore, the conclusion was that the sun was undergoing apparently severe nuclear reaction transformations in its fuel chains and cycles. Other affected spectrographs were for Nitrogen, and Carbon, which also showed increased intensity and magnitude in electromagnetic spectral signatures.

All media organizations and agencies switched programming in order to begin reporting on these events. Many had agreements with particular scientists who made themselves ready and available for teleconferences and on-site analyses so that all whole-world populations could keep abreast of continuing developments for emergency response preparedness and active readiness.

The sun increased in brightness and intensity of radio-magnetism. Observatory machines with radio, sonar and radar wavelengths displayed extreme activity as if the sun was a roaring and blasting, raging and exploding inferno, spewing out long plumes of radioactive electro-flames.

One machine began to "smoke" at Marshall Space Flight Center as the speakers's paper diaphragm tore in half. The machine was immediately disconnected, the data retrieved and the fire extinguished.

Earth atmosphere "felt heavy" to human beings, many of whom began to develop difficulty in breathing. Rather, the air appeared to have "thinned," as if suffering decreases in Oxygen density, even at the surface of the Earth. Many people fainted, while others had to sit or lay down in order to avoid falling. Hospital and health centers, clinics and medical agencies, all, have received special training in expectation of unusual or uncommon biological, immunological and metabolic symptoms due to gravi-metric and radio-magnetic occurrences. Certain population groups like babies, infants and the elderly, the already ill, and those who had undergone surgery, all, received immediate attention in order to prevent massive death occurrences.

The chain reactions from Hydrogen to Helium appeared to have climaxed in Helium "burning" continuously. If there were still Hydrogen atoms available, they had undergone "cooling" or could not ignite for continuum chain cycling with Helium, but would ignite to fuse again as pressure increased and heat maximized in intensity. Radio-magnetic field strength was so strong that the sun appeared to have "flattened" slightly at the poles, but with an unusual change—the sun's "sides" from our perspective on the Earth appeared to be "oscillating" back and forth between "flattening" or "contracting" inwards and "extending" or "bulging" outwardly. The sun's magnetic field was circulating around its spherical periphery while "counter-binding" dipolar magnetic energies.

"Side-vectored magnetic field polarity! Wow! That is amazing!" exclaimed Colbert to Stewart.

"Colbert," muttered Stewart as if stunned by the image on the screen facing him. "Do you know what these two things are that we're looking at, at each pole of the sun? These are 'coronal holes,' the biggest we've ever seen. The one at the North Pole appears to be much bigger than that at the South Pole. They are so huge and 'dark.'"

"Stewart," replied Colbert, "Don't these things occur usually after the sun has reached 'sunspot maximum,' that is, some years after 'polarity change'?"

"Yes," added Stewart, "At least, that's the theory. But these times are really different, given what has been happening. It looks like during 'pangs of sun death,' if that's what's really taking place, all these radio-magnetic processes at the core and within convection regions, have interchanged intensity, magnitude, sequence and periodicity. There appears to be total but contained chaos within the photosphere! What a hellish place that would be! No wonder the sun is so 'mad'!"

"That means, the sun is undergoing extreme conditions of potential cooling which might throw all variables, parameters and determinants 'out of kilter,'" continued Colbert.

It appeared that "contracting" and "extending" was occurring at all places along the sun's peripheral circumference region, due to explosive coronal emissions that were then immediately suctioned back inwardly into the solar photosphere—where sun spots, flares, winds and plasma residues intermingled to "toggle" temperature and pressure parameters, from one extreme to the other. The omni-directional pluri-potent magnetic fields packets had gained so much excitation as to have extended their heat flux from the sun's center, radially outwards, being stopped only by extreme magnetic field compressive activities that were now "running" all along the sun's outer perimeter. For, the sun's implosive forces were as "omni-vector potent" as the expanding thermal plasma-particulates driving its explosive momentum.

As particles transmuted and trans-morphed themselves in the extreme core plasma heat and pressure, radiative energy greatly increased and the core began to recycle Nitrogen and Carbon, the latter two heavier elements serving therefore as "buffers" interfacing with plasma fusion fire dynamics to stabilize the reaction chain. Even neutrinos were "bouncing to and fro" in their explosive attempts to escape from the radio-magnetic sphere.

But at the same time, the sun underwent severe "cooling" in comparison to the great and extreme heat intensities that are necessary for its Hydrogen-Helium fusion cycles. These cycles had lost their predictable patterns in the conflagration now taking place within the sun's horrific inferno. As these heavier protonic nuclei lost their electrons and neutrons to the core fuel combustion, the whole sun appeared to "shrivel" and "wrinkle" in texture, appearance and composition, "flirting with an ellipsoidal form," due to radiation mass seeking to escape into cosmic space—as if something was missing;—heat loss could not continue in this manner lest the transition from "yellow dwarf" to "red giant" be unwarrantedly accelerated beyond any solar ability to regain the Hydrogen-Helium cycle for continuum Qualitative Conservation of elemental and molecular spectrum energetics. The spectral ejections were massive, but not as extensive as would occur if the "red giant" stage had been reached. Nitrogen and Carbon cycling as atomic energy appeared to have effected a modicum of stability in thermonuclear reactions.

The Earth was gyrating, convulsing, contracting, expanding, tossing, heaping, collapsing and rebounding, but its magnetic field, holding fast—as if the nausea experienced by human beings was also being "felt" by its elements and components, in ways that were totally different from shocks common during earthquake activities, as it struggled to sustain synchronized equilibrium.

Conservation of energy and of mass-momentum is paramount in cosmic systems processes, meaning momentum equilibrium could be lost in the presence of extremely disproportionate entropic entanglements. Conservation of equilibrium is necessary for thermodynamic operational consistency.

Tectonic activity was minimal at this time, but the oceans were raging in tempest, the atmosphere was devoured by extremely strong winds, and the landmass was being "slowly swayed" in all directions, up and down and from side to side, vectoring with different magnitudes and strength, in multi-causal velocities, within the spherical Earth, as "buffered" by the interface of the landmass with the atmosphere and hydrosphere, a process which then exacerbated and intensified the convective vortex-creating momentum gained by winds and ocean currents, hence, disrupting "immuno-metabolic" and "neuro-electrolytic" processes in the human organism.

"O, I can't stand it anymore," said one man working at a network television station. "My stomach is in turmoil and I can't puke. I have not eaten since this morning. I feel 'light-headed,' and as if I had spent the whole night drinking alcohol and have not slept a wink for 24 hours. But as you all know this is not the case. This is the strangest feeling—like 'motion sickness' but worse."

"Here, have some crackers," replied, one of his co-workers. "It's got salt and that will help stabilize your bile system. Drink some water too."

"It's the solar system, the sun, the Earth, the Moon and the galaxies—it's the whole universe. We're all like that, more or less. Some of us 'fake it' better than others. We just have to 'brace ourselves' for a while and hopefully things will return to normal. God help us," added another worker.

By then, it was 9:00am on that Wednesday. Schools were already functioning, but under heavy duress. Many teachers and students were ill from "motion sickness," so to speak, for lack of a better word. In the streets, automobile traffic had temporarily stopped, with many drivers deciding to park their cars, vans, and trucks temporarily, until the Earth showed "signs of calming down."

On radio and television, it was explained that human ears have little bones called "otolith" that transmit vibrations onto the inner ear in order to help give us a sense of normal equilibrium as we move, or balance as we travel and run, as the inner ear also is filled with small hair "follicles" and electrostatic fluid through which signals pass onto the cortex for processing. Gravitational and radiometric effects—radiation, gravity, electro-magnetism, pressure and temperature variables etc . . . —on our organism, have disrupted this "sense of equilibrium" that we commonly and usually "feel" but seldom pay attention to, as we go about our normal business in life.

Students at the high school in Illinois remembered that flock of geese that flew right into the central structure of the high school building to crash unto the ground as they were killed instantly. The radio and television commenting scientists explained that animals also possessed analogous ear structures that allowed them to navigate on the Earth while maintaining their equilibrium in all their existence motions.

School superintendents, along with district principals, were considering releasing all teachers and students from class attendance for the day, due to circumstances that negatively affected everyone's ability to perform, but were waiting to hear from other constituencies in order to minimize panic and "bottlenecks" in the streets of towns and cities taking similar actions.

The President of the United States had called all the media for a press conference, flanked and encompassed by many scientists and other civilian people from organizations so diverse as to signify a momentous occasion. With much alacrity, he summoned all courage to exhort Americans and others to show patience and forbearance in these times of solar system turbulence. He indicated, if possible transportation was forthcoming, it would be a good idea for students to remain in a familiar environment and space, such as at home, where conditions were familiar and events more or less routine, in order to facilitate care in case of illness or discomfort. He encouraged family doctors to remain on call, if they themselves were not too ill, and to network with their fellow professionals in order to have some gauge on the number of practicing physicians caring for our population in these times of distress.

The school districts had reported they had to "hire extra help" to clean up classrooms from student nausea and teacher illnesses due to radio-magnetic related effects and solar gravi-metrics events affecting all parties at school. Gravity-sensitive human senses had responded in that manner to protect the organism from shock or other more severe damage. For the sun's radio-magnetic and gravitational "push and pull" contortions against the Earth and the Moon had gravely subjected human beings to the most counterintuitive motions ever felt since the time of Creation.

It felt as if they were all in a big boat on the oceans, surviving the greatest storms ever seen in the universe, while riding on swings. The sun appeared to be "playing seesaw" with the planets as each was being affected in ways corresponding to the types of space-bodies they were. Here on Earth, there was life, however, Human Life, which made all the difference in this universe that seemed to be "out of control."

The President, therefore, agreed with county superintendents that students and teachers should be sent on their way home, once safe and secure transport was possible. However, he cautioned that, other institutions, such as fire departments, hospitals and other emergency response entities ought to be very prudent in applying these standards to themselves; that they should rather, in consultation with urban legislatures and state assemblies confer on public acts that would facilitate their access to medical care and appropriate medications designed to secure their "remaining on the job."

He continued by congratulating every one for their dedicated service to the nation and appreciating the responsible care local and state agencies had been taking in affording the greatest latitude in self-government for the population in their respective connections to the handling of preparedness, readiness, and active emergency response.

*   *   *

20

THE SUN HAD NOT totally returned to regular radio-gra-vi-metric equilibrium—the core, the photosphere, the convective sphere and the magnetosphere were all in coun-ter-momentum force electro-motive perturbations and con-volutions. Not only was the magnetic field circling the solar circumference, but the sun's rotational motions around the galaxy appeared to have slightly accelerated, which might explain transformative changes in Earth-sensitive gravity effects. Solar core densities had greatly increased due to the cycling of heavier elements which were also responsible for relative "cooling" of the solar dynamo, increasing the intensity of the magnetic field towards the more extreme compressive spectrum of momentum forces.

There was a final polar contortion during which the "coronal holes" glowed with brightness to then be engulfed within the poles as if violently pulled-in, as the sun quaked and shook for a few minutes of re-adjustment and re-cali-bration of dynamo and convective temperature flux and pressure parameters. These solar prominences had cata-pulted various particulate flows towards the solar equator to intermix with the oncoming plasma derivatives headed towards coronal processing, which triggered the greatest "tumultuous agitation" in the convection zone.

In addition, solar luminosity had decreased to such an extent as to have triggered the lighting of "photosensitive" cells and bulbs on the Earth even within the morning hours, but core-convection zone counter-reactions within the sun evoking huge "suction vortices" then caused the surface to re-intensify in brightness by mid-afternoon.

These "oscillations" comparable to a pendulum that had been "wrung up" by a tremendous compressive force, but as to distend more or less rapidly, caused diverse extra-range effects that superseded temperature and pressure gradients conducive to Earth system equilibrium and geo-eco-stability.

The sun's newly acquired cosmic rotational motion increase had caused a contraction in the "spring plane" or "rubber band plane" that tied all planets to its gravitationally attractive binding energies, as all planets were "forced inwards" within and towards the center of the sun's heliospheric mass, to then be slowly repelled by solar system center of field energies. Jupiter, the planet with the greatest mass, had reacted violently with electro-motive inwardly directed atmospheric fires whereas Mercury had exemplified a "double shift" at both aphelion and perihelion.

As all planets reacted to this new phenomenon, the whole solar system appeared to "behave" as a huge "bowl of soup" wherein all ingredients in the "hot cosmic fluid" were "randomly" waving, rolling, tossing and heaving, and interacting per extreme solar magnetic field variables unknown to humans thus far in our living existence upon the Earth.

Hydrogen appeared to have regained its cyclical periodicity into Helium reactions as the core began to increase in plasma fusion temperatures and pressures more akin to regularly combusting star systems. The cycling of heavier Nitrogen atoms had contributed to solar "cooling" while Carbon cycles had induced reactive stability that facilitated the reinstatement of dynamic equilibrium in a time span that covered a period of 5 hours during which the Earth had awkwardly "tumbled" from being "tossed over and under" in all possible directions within spherical angular momentum degree variations that engendered numerous cases of destruction, illness, discomfort and death thereupon.

Many activities were disrupted and others had to be discontinued until the next day, as perturbations, emanations, reverberations and resonances from solar gravi-radio spectral emissions had subsided to the point of assuming a "calm period" had begun. Earth oceans had processed tremendous amounts of heat as the atmosphere had absorbed tremendous amounts of radiation from magnetosphere contact with field-transported plasma winds that penetrated into the ionosphere. Atmospheric layers or strata were very active in processing these gravi-metric differentials that had accounted for changes in initial conditions of operational functional equilibrium.

Oceans were still greatly animated by tidal activity, having absorbed the greater amount of landmass swings and sways which had aggravated human equilibrium sensitive organs that helped them maintain balance during operational motion or working movement.

Many ships that were at sea at the time had to lay anchor, lest they be suctioned downwards into the oceans by massive vortices created by gravitational differentials which they could not overcome with their engine power. Others that remained in port suffered great structural damage as well as bent screws and distorted drive shafts as they were "banged" against port frames which also received severe destructive blows.

Earth landmass, hydrosphere and atmosphere had undergone so many turbulent periods of destabilization that many ecological processes and planetary events will remain in apparent "disorder" for a while, that is, until conservation determinants proceeded to naturally repair the Earth, as had usually occurred after a volcanic eruption, earthquake, hurricane or tornado had "gone through."

*   *   *

www.ingramcontent.com/pod-product-compliance
Lightning Source LLC
Chambersburg PA
CBHW051141020726
47501CB00005B/1625